Theory of Photon Acceleration

Series in Plasma Physics

Series Editors:

Professor Peter Stott, CEA Caderache, France
Professor Hans Wilhelmsson, Chalmers University of Technology, Sweden

Other books in the series

An Introduction to Alfvén Waves
R Cross

Transport and Structural Formation in Plasmas
K Itoh, S-I Itoh and A Fukuyama

Tokamak Plasma: a Complex Physical System
B B Kadomtsev

Electromagnetic Instabilities in Inhomogeneous Plasma
A B Mikhailovskii

Instabilities in a Confined Plasma
A B Mikhailovskii

Physics of Intense Beams in Plasma
M V Nezlin

The Plasma Boundary of Magnetic Fusion Devices
P C Stangeby

Collective Modes in Inhomogeneous Plasma
J Weiland

Forthcoming titles in the series

Plasma Physics via Computer Simulation, 2nd Edition
C K Birdsall and A B Langdon

Nonliner Instabilities in Plasmas and Hydrodynamics
S S Moiseev, V G Pungin, and V N Oraevsky

Laser-Aided Diagnostics of Plasmas and Gases
K Muraoka and M Maeda

Inertial Confinement Fusion
S Pfalzner

Introduction to Dusty Plasma Physics
P K Shukla and N Rao

Series in Plasma Physics

Theory of Photon Acceleration

J T Mendonça

Instituto Superior Técnico, Lisbon

CRC Press is an imprint of the
Taylor & Francis Group, an **informa** business

CRC Press
Taylor & Francis Group
6000 Broken Sound Parkway NW, Suite 300
Boca Raton, FL 33487-2742

First issued in paperback 2019

© 2001 by Taylor & Francis Group, LLC
CRC Press is an imprint of Taylor & Francis Group, an Informa business

No claim to original U.S. Government works

ISBN-13: 978-0-7503-0711-6 (hbk)
ISBN-13: 978-0-367-39784-5 (pbk)

Visit the Taylor & Francis Web site at
http://www.taylorandfrancis.com

and the CRC Press Web site at
http://www.crcpress.com

British Library Cataloguing-in-Publication Data

A catalogue record for this book is available from the British Library.

Library of Congress Cataloging-in-Publication Data are available

Cover Design: Victoria Le Billon

Typeset in TEX using the IOP Bookmaker Macros

This figure illustrates the first experimental results obtained by the IST group on photon acceleration by a relativistic ionization front. From left to right, we can see: the spectrum of the initial laser pulse, the up-shifted spectrum, in counter-propagation and the up-shifted spectrum in co-propagation. These spectra are artificially coloured, with the colours exactly corresponding to the observed frequencies.

Onda que, enrolada, tornas,
Pequena, ao mar que te trouxe
E ao recuar te transtornas
Como se o mar nada fosse
Fernando Pessoa

To my children:
Dina, Joana and Pedro

Contents

Acknowledgments		**xi**
1	**Introduction**	**1**
	1.1 Definition of the concept	1
	1.2 Historical background	3
	1.3 Description of the contents	5
2	**Photon ray theory**	**8**
	2.1 Geometric optics	8
	2.2 Space and time refraction	11
	2.2.1 Refraction	12
	2.2.2 Time refraction	13
	2.2.3 Space–time refraction	15
	2.3 Generalized Snell's law	17
	2.4 Photon effective mass	22
	2.5 Covariant formulation	27
3	**Photon dynamics**	**31**
	3.1 Ionization fronts	31
	3.2 Accelerated fronts	39
	3.3 Photon trapping	43
	3.3.1 Generation of laser wakefields	43
	3.3.2 Nonlinear photon resonance	44
	3.3.3 Covariant formulation	49
	3.4 Stochastic photon acceleration	51
	3.4.1 Motion in two wakefields	52
	3.4.2 Photon discrete mapping	54
	3.5 Photon Fermi acceleration	57
	3.6 Magnetoplasmas and other optical media	63
4	**Photon kinetic theory**	**67**
	4.1 Klimontovich equation for photons	68
	4.2 Wigner–Moyal equation for electromagnetic radiation	70
	4.2.1 Non-dispersive medium	70
	4.2.2 Dispersive medium	74

4.3   Photon distributions   75
    4.3.1   Uniform and non-dispersive medium   76
    4.3.2   Uniform and dispersive medium   77
    4.3.3   Pulse chirp   78
    4.3.4   Non-stationary medium   81
    4.3.5   Self-blueshift   82
4.4   Photon fluid equations   83
4.5   Self-phase modulation   87
    4.5.1   Optical theory   88
    4.5.2   Kinetic theory   90

5   Photon equivalent charge   97
5.1   Derivation of the equivalent charge   97
5.2   Photon ondulator   103
5.3   Photon transition radiation   105
5.4   Photon Landau damping   108
5.5   Photon beam plasma instabilities   113
5.6   Equivalent dipole in an optical fibre   115

6   Full wave theory   123
6.1   Space and time reflection   123
    6.1.1   Reflection and refraction   123
    6.1.2   Time reflection   125
6.2   Generalized Fresnel formulae   126
6.3   Magnetic mode   128
6.4   Dark source   133

7   Non-stationary processes in a cavity   140
7.1   Linear mode coupling theory   140
7.2   Flash ionization in a cavity   143
7.3   Ionization front in a cavity   146
7.4   Electron beam in a cavity   149
7.5   Fermi acceleration in a cavity   152

8   Quantum theory of photon acceleration   157
8.1   Quantization of the electromagnetic field   157
    8.1.1   Quantization in a dielectric medium   157
    8.1.2   Quantization in a plasma   161
8.2   Time refraction   165
    8.2.1   Operator transformations   165
    8.2.2   Symmetric Fock states   167
    8.2.3   Probability for time reflection   170
    8.2.4   Conservation relations   172
8.3   Quantum theory of diffraction   173

9   New developments   **177**
   9.1   Neutrino–plasma physics   177
   9.2   Photons in a gravitational field   183
          9.2.1   Gravitational redshift   183
          9.2.2   Gravitational lens   186
          9.2.3   Interaction of photons with gravitational waves   187
          9.2.4   Other metric solutions   191
   9.3   Mean field acceleration processes   193

Appendix   Derivation of the Wigner–Moyal equation   **195**
   A.1   Non-dispersive media   195
   A.2   Dispersive media   198

References   **203**

Glossary   **209**

Index   **215**

Acknowledgments

This book results from a very fruitful collaboration, at both local and international levels. First of all, I would like to thank Robert Bingham and Padma K Shukla for their contagious enthusiasm about science, and for their collaboration in several different subjects. I also would like to thank Nodar L Tsintsadze, for his active and creative collaboration during his long stay in Lisboa. I extend my thanks to my many other collaborators in this field.

At the local level, I would like to thank my students, with particular emphasis to Luis O Silva, who efficiently transformed some of my vague suggestions into coherent scientific papers. I am also gratefull to Joao M Dias, Nelson Lopes, Gonçalo Figueira, Carla C Rosa, Helder Crespo, Madalena Eloi and Ricardo Fonseca, who by their successful experiments and numerical simulations helped me to get a better understanding of this field. I would also like to refer to the more recent collaboration with Ariel Guerreiro and my colleague Ana M Martins on the quantum theory. J M Dias, G Figueira and J A Rodrigues prepared some of the figures in this book.

My acknowledgments extend to my colleagues Armando Brinca and Filipe Romeiras, who looked carefully at a first version of this book and provided very useful suggestions and comments.

J T Mendonça
July 1999

Chapter 1

Introduction

We propose in this book to give a simple and accurate theoretical description of photon acceleration, and of related new concepts such as the effective photon mass, the equivalent photon charge or the photon Landau damping.

We also introduce, for the first time, the concepts of time reflection and time refraction, which arise very naturally from the theory of wave propagation in non-stationary media. Even if some of these concepts seem quite exotic, they nevertheless result from a natural extension of the classical (and quantum) electrodynamics to the cases of very fast processes, such as those associated with the physics of ultra-short and intense laser pulses.

This book may be of relevance to research in the fields of intense laser–matter interactions, nonlinear optics and plasma physics. Its content may also help to develop novel accelerators based on laser–plasma interactions, new radiation sources, or even to establish new models for astrophysical objects.

1.1 Definition of the concept

The concept of photon acceleration appeared quite recently in plasma physics. It is a simple and general concept associated with electromagnetic wave propagation, and can be used to describe a large number of effects occurring not only in plasmas but also in other optical media. Photon acceleration is so simple that it could be considered a trivial concept, if it were not a subtle one.

Let us first try to define the concept. The best way to do it is to establish a comparison between this and a few other well-known concepts, such as with refraction. For instance, photon acceleration can be seen as a space–time refraction.

Everybody knows that refraction is the change of direction suffered by a light beam when it crosses the boundary between two optical media. In more technical terms we can say that the wavevector associated with this light beam changes, because the properties of the optical medium vary in space.

We can imagine a symmetric situation where the properties of the optical medium are constant in space but vary in time. Now the light wavevector remains constant (the usual refraction does not occur here) but the light frequency changes. This effect, which is as universal as the usual refraction, can be called time refraction. A more general situation can also occur, where the optical medium changes in both space and time and the resulting space–time refraction effect coincides with what is now commonly called photon acceleration.

Another natural comparison can be established with the nonlinear wave processes, because photon acceleration is likewise responsible for the transfer of energy from one region of the electromagnetic wave spectrum to another. The main differences are that photon acceleration is a non-resonant wave process, because it can allow for the transfer of electromagnetic energy from one region of the spectrum to an arbitrarily different one, with no selection rules.

In this sense it contrasts with the well-known resonant wave coupling processes, like Raman and Brillouin scattering, harmonic generation or other three- or four-wave mixing processes, where spectral energy transfer is dictated by well-defined conservation laws. We can still say that photon acceleration is a wave coupling process, but this process is mainly associated with the linear properties of the space and time varying optical medium where the wavepackets propagate, and the resulting frequency shift can vary in a continuous way.

Because this is essentially a linear effect it will affect every photon in the medium. This contrasts, not only with the nonlinear wave mixing processes, but also with the wave–particle interaction processes, such as the well-known Compton and Rayleigh scattering, which only affect a small fraction of the incident photons. We can say that the total cross section of photon acceleration is equal to 1, in contrast with the extremely low values of the Compton or Rayleigh scattering cross sections. Such a sharp difference is due to the fact that Compton and Rayleigh scattering are single particle effects, where the photons interact with only one (free or bounded) electron, while photon acceleration is essentially a collective process, where the photons interact with all the charged particles of the background medium.

This means that photon acceleration can only be observed in a dense medium, with a large number of particles at the incident photon wavelength scale. For instance, in a very low density plasma, an abrupt transition from photon acceleration to Compton scattering will eventually occur for decreasing electron densities. This may have important implications, for instance, in astrophysical problems.

The concept of photon acceleration is also a useful instrument to explore the analogy of photons with other more conventional particles such as electrons, protons or neutrons. Using physical intuition, we can say that the photons can be accelerated because they have an effective mass (except in a vacuum).

In a plasma, the photon mass is simply related with the electron plasma frequency, and, in a general optical medium, this effective mass is a consequence of the linear polarizability of the medium.

This photon effective mass is, in essence, a linear property, but the particle-like aspects of the electromagnetic radiation, associated with the concept of the photon, can also be extended in order to include the medium nonlinearities.

The main nonlinear property of photons is their equivalent electric charge, which results from radiation pressure, or ponderomotive force effects. In a plasma, this ponderomotive force tends to push the electrons out of the regions with a larger content of electromagnetic wave energy. Instead, in an optical fibre or any similar optical medium, the nonlinear second-order susceptibility leads to the appearance of an equivalent electric dipole. Because these equivalent charge distributions, monopole or dipole charges according to the medium, move at very fast speeds, they can act as relativistic charged particles (electrons for instance) and can eventually radiate Cherenkov, transition or bremsstrahlung radiation.

These and similar effects have recently been explored in plasmas and in nonlinear optics, especially in problems related to ultra-short laser pulse propagation [3, 23], or to new particle accelerator concepts and new sources of radiation [42]. The concept of photon acceleration can then be seen as a kind of new theoretical paradigm, in the sense of Kuhn [51], capable of integrating in a unified new perspective, a large variety of new or already known effects associated with electromagnetic radiation.

Furthermore, we can easily extend this concept to other fields, for instance to acoustics, where phonon acceleration can also be considered [1]. This photon phenomenology can also be extended to the physics of neutrinos in a plasma (or in neutral dense matter), if we replace the electromagnetic coupling between the bound or free electrons with the photons by the weak coupling between the electrons and the neutrino field [106].

Taking an even more general perspective, we can also say that photon acceleration is a particular example of a mean field acceleration process, which can act through any of the physical interaction forces (electromagnetic, weak, gravitational or even the strong interactions). This means that, for instance, particles usually considered as having no electric charge can efficiently be accelerated by an appropriate background field.

This has been known for many centuries (since the invention of slingshots), but was nearly ignored by the builders of particle accelerators of our days. It also means that particles with no bare electric charge can polarize the background medium, and become 'dressed' particles with induced electric charge, therefore behaving as if they were charged particles.

1.2 Historical background

Let us now give a short historical account of this concept. The basic equations necessary for the description of photon acceleration have been known for many years, even if their explicit meaning has only recently been understood. This is due to the existence of a kind of conceptual barrier, which prevented formally simple jumps in the theory to take place, which could provide a good example of what Bachelard would call an *obstacle epistemologique* [6].

The history of the photon concept is actually rich in these kinds of conceptual barrier, and the better known example is the 30 year gap between the definition by Einstein of the photon as a quantum of light, with energy proportional to the frequency ω, and the acceptance that such a particle would also have a momentum, proportional to the wavevector $\vec{k}$, introduced in the theory of Compton scattering [86].

It is therefore very difficult to have a clear historical view and to find out when the concept of photon acceleration clearly emerged from the already existing equations. The assumed subjective account of the author of the photon acceleration story will be proposed, accepting that other and eventually better and less biased views are also possible.

One of the first papers which we can directly relate to photon acceleration in plasma physics was published by Semenova in 1967 [97] and concerns the frequency up-shift of an electromagnetic wave interacting with a moving ionization front. At this stage, the problem could be seen as an extension of the old problem of wave reflection by a relativistic mirror, if the mirror is replaced by a surface of discontinuity between the neutral gas and the ionized gas (or plasma).

The same problem was later considered in greater detail by Lampe *et al* in 1978 [53]. In these two papers, the mechanism responsible for the ionization front was not explicitly discussed. But more recently it became aparent that relativistic ionization fronts could be produced in a laboratory by photoionization of the atoms of a neutral gas by an intense laser pulse.

A closely related, but qualitatively different, mechanism for photon acceleration was considered by Mendonça in 1979 [63] where the ionization front was replaced by a moving nonlinear perturbation of the refractive index, caused by a strong electromagnetic pulse. In this work it was shown that the frequency up-shift is an adiabatic process occurring not only at reflection as previously considered, by also at transmission.

At that time, experiments like those of Granatstein *et al* [36, 87], on microwave frequency up-shift from the centimetric to the milimetric wave range, when reflected inside a waveguide by a relativistic electron beam, were exploring the relativistic mirror concept. The idea behind that theoretical work was to replace the moving particles by moving field perturbations, easier, in principle, to be excited in the laboratory.

In a more recent work produced in the context of laser fusion research, Wilks *et al* [118] considered the interaction of photons with plasma wakefield perturbations generated by an intense laser pulse. Using numerical simulations, they were able to observe the same kind of adiabatic frequency up-shift along the plasma. This work also introduced for the first time the name of *photon acceleration*, which was rapidly adopted by the plasma physics community and stimulated an intense theoretical activity on this subject.

A related microwave experiment by Joshi *et al*, in 1992 [95] was able to show that the frequency of microwave radiation contained in a cavity can be up-shifted to give a broadband spectrum, in the presence of an ionization front produced by an ultraviolet laser pulse. These results provided the first clear indication that the photon acceleration mechanism was possibly taking place.

In the optical domain, the observation of a self-produced frequency up-shift of intense laser pulses, creating a dense plasma when they are focused in a neutral gas region, and the measured up-shifts [120], were also pointing to the physical reality of the new concept and giving credit to the emerging theory. Very recently, the first two-dimensional optical experiments carried out by our group [21], where a probe laser beam was going through a relativistic ionization front in both co- and counter-propagation, were able to demonstrate, beyond any reasonable doubt, the existence of photon acceleration and to provide an accurate quantitative test of the theory.

Actually, the spectral changes of laser beams by ionization of a neutral gas were reported as early as 1974 by Yablonovich [122]. In these pioneering experiments, the spectrum of a CO_2 laser pulse was strongly broadened and slightly up-shifted, when the laser beam was focused inside an optical cavity and ionization of the neutral gas inside the cavity was produced. This effect is now called flash ionization for reasons that will become apparent later, and it can also be considered as a particular and limiting case of the photon acceleration processes.

In parallel with this work in plasma physics, and with almost completely mutual ignorance, following both theoretical and experimental approaches clearly independently, research on a very similar class of effects had been taking place in nonlinear optics since the early seventies. This work mainly concentrated on the concept of phase modulation (including self-, induced and cross phase modulation), and was able to prove both by theory and by experiments, that laser pulses with a very large spectrum (called the supercontinuum radiation source) can be produced. This is well documented in the book recently edited by Alfano [3].

As we will see in the present work, the theory of photon acceleration as developed in plasma physics is also able to explain the phase modulation effects, when we adapt it to the optical domain. This provides another proof of the interest and generality of the concept of photon acceleration. We will attempt in this work to bridge the gap between the two scientific communities and between the two distinct theoretical views.

1.3 Description of the contents

Four different theoretical approaches to photon acceleration will be considered in this work: (1) single photon trajectories, (2) photon kinetic theory, (3) classical full wave models and (4) quantum theory.

The first two chapters will be devoted to the study of single photon equations (also called ray equations), derived in the frame of geometric optics. This is the simplest possible theoretical approach, which has several advantages over more accurate methods. Due to its formal simplicity, we can apply it to describe, with great detail and very good accuracy, various physical configurations where photon acceleration occurs. These are ionization fronts (with arbitrary shapes and velocities), relativistic plasma waves or wakefields, moving nonlinearities and flash ionization processes.

It can also be shown that stochastic photon acceleration is possible in several physical situations, leading to the transformation of monocromatic radiation into white light. An interesting example of stochastic photon behaviour is provided by the well-known Fermi acceleration process [29], applied here to photons, which can be easily described with the aid of single photon equations.

Apart from its simplicity and generality, photon ray equations are formally very similar to the equations of motion of a material particle. This means that photon acceleration happens to be quite similar to electron or proton acceleration by electromagnetic fields, even if the nature of the forces acting on the photons is not the same. For instance, acceleration and trapping of electrons and photons can equally occur in the field of an electron plasma wave.

Chapter 2 deals with the basic concepts of this single photon or ray theory, as applied to a generic space- and time-varying optical medium. The concept of space–time refraction is introduced, the generalized Snell's laws are derived and the ray-tracing equations are stated in their Hamiltonian, Lagrangian and covariant forms.

Chapter 3 deals with the basic properties of photon dynamics, illustrated with examples taken from plasma physics, with revelance to laser–plasma interaction problems. Extension to other optical media, and to nonlinear optical configurations will also be discussed.

This single photon theory is simple and powerful, but it can only provide a rough description of the laser or other electromagnetic wavepackets evolving in non-stationary media. However, an extension of this single particle description to the kinetic theory of a photon gas is relatively straightforward and can lead to new and surprising effects such as photon Landau damping [14]. This is similar to the well-known electron Landau damping [54].

Such a kinetic theory is developed in chapter 4 and gives a much better description of the space–time evolution of a broadband electromagnetic wave spectrum. This is particularly important for ultra-short laser pulse propagation. In particular, self-phase modulation of a laser pulse, propagating in a nonlinear optical medium, and the role played by the phase of the laser field in this process, will be discussed.

Chapter 5 is devoted to the discussion of the equivalent electric charge of photons in a plasma, and of the equivalent electric dipole of photons in an optical fibre. We will also discuss the new radiation processes associated with these charge distributions, such as photon ondulator radiation, photon transition radiation or photon bremsstrahlung.

The geometric optics approximation, in its single photon and kinetic versions, provides a very accurate theoretical description for a wide range of different physical configurations. Even very fast time events, occurring on a timescale of a few tens of femtoseconds, can still be considered as slow processes in the optical domain and stay within the range of validity of this theory. But, in several situations, a more accurate theoretical approach is needed, in order to account for partial reflection, for specific phase effects, or for arbitrarily fast time processes.

We are then led to the full wave treatment of photon acceleration, which is presented in chapters 6 and 7. Most of the problems discussed in previous chapters are reviewed with this more exact approach and a comparison is made with the single photon theory when possible. New aspects of photon acceleration can now be studied, such as the generation of a magnetic mode, the multiple mode coupling or the theory of the dark source which describes the possibility of accelerating photons initially having zero energy.

In chapter 8 we show that a quantum description of photon acceleration is also possible. We will try to establish in solid grounds the theory of time refraction, which is the basic mechanism of photon acceleration.

The quantum Fresnel formulae for the field operators will be derived. We will also show that time refraction always leads to the creation of photon pairs, coming out of the vacuum. More work is still in progress in this area.

Finally, chapter 9 is devoted to new theoretical developments. Here, the photon acceleration theory is extended in a quite natural way to cover new physical problems, which correspond to other examples of the mean field acceleration process. These new problems are clearly more controversial than those covered in the first eight chapters, but they are also very important and intellectually very stimulating.

Two of these examples are briefly discussed. The first one concerns collective neutrino plasma interaction processes, which were first explicitly formulated by Bingham *et al* in 1994 [13] and have recently received considerable attention in the literature. The analogies between the photon and the neutrino interaction with a background plasma will be established. The second example will be the interaction of photons with a gravitational field and the possibility of coupling between electromagnetic and gravitational waves. One of the consequences of such an interaction is the occurrence of photon acceleration in a vacuum by gravitational waves.

Chapter 2

Photon ray theory

It is well known that the wave–particle dualism for the electromagnetic radiation can be described in purely classical terms. The wave behaviour is described by Maxwell's equations and the particle behaviour is described by geometric optics. This contrasts with other particles and fields where the particle behaviour is described by classical mechanics and the wave behaviour by quantum mechanics.

Geometric optics is a well-known and widely used approximation of the exact electromagnetic theory and it is presented in several textbooks [15, 16]. We will first use the geometric optics description of electromagnetic wavepackets propagating in a medium. These wavepackets can be viewed as classical particles and can be assimilated to photons. We will then apply the word 'photon' in the classical sense, as the analogue of an electromagnetic wavepacket. A single photon can be used to represent the mean properties of a given wavepacket. The photon velocity will then be equal to the group velocity of the wavepacket.

This single photon approach will be used in the present and the next chapters. A more accurate description of a wavepacket can still be given in the geometric optics approximation, by using a bunch of photons, instead of a single one. The study of such a bunch will give us information on the internal spectral content of the wavepacket. This will be discussed in chapter 4. The use of the photon concept in a quantum context will be postponed until chapter 8.

2.1 Geometric optics

It is well known that a wave is a space–time periodic event, where the time periodicity is characterized by the angular frequency ω and the space periodicity as well as the direction of propagation are characterized by the wavevector $\vec{k}$. In particular, an electromagnetic wave can be described by an electric field of the form

$$\vec{E}(\vec{r}, t) = \vec{E}_0 \exp i\{\vec{k} \cdot \vec{r} - \omega t\} \tag{2.1}$$

where $\vec{E}_0$ is the wave field amplitude, $\vec{r}$ is the position and t the time.

In a stationary and uniform medium the frequency and the wavevector are constants, but they are not independent from each other. Instead, they are related by a well-defined expression (at least for low amplitude waves) known as the dispersion relation. In a vacuum, the dispersion relation is simply given by $\omega = kc$, where the constant c is the speed of light in a vacuum and k is the absolute value of the wavevector (also known as the wavenumber).

In a medium, the dispersion relation becomes

$$\omega = \frac{kc}{n} = \frac{kc}{\sqrt{\epsilon}} = \frac{kc}{\sqrt{1 + \chi}} \tag{2.2}$$

where n is the refractive index of the medium, ϵ its dielectric constant and χ its susceptibility. In a lossless medium these quantities are real, and in dispersive media they are functions of the frequency ω and the wavevector $\vec{k}$.

In particular, for high frequency transverse electromagnetic waves propagating in an isotropic plasma [82, 108], we have $\epsilon = 1 - (\omega_p/\omega)^2$, where ω_p is the electron plasma frequency. It is related to the electron density n_e by the expression: $\omega_p^2 = e^2 n_e/\epsilon_0 m$, where e and m are the electron charge and mass, and ϵ_0 is the vacuum permittivity.

From equation (2.2) we can then have a dispersion relation of the form

$$\omega = \sqrt{k^2 c^2 + \omega_p^2}. \tag{2.3}$$

A similar equation is also valid for electromagnetic waves propagating in a waveguide [25]. The plasma frequency is now replaced by a cut-off frequency ω_0 depending on the field configuration and on the waveguide geometry.

In the most general case, however, the wave dispersion relation is a complicated expression of ω and $\vec{k}$, and such simple and explicit expressions for the frequency cannot be established. It is then preferable to state it implicitly as

$$R(\omega, \vec{k}) = 0. \tag{2.4}$$

The above description of wave propagation is only valid for uniform and stationary media. Let us now assume that propagation is taking place in a non-uniform and non-stationary medium. If the space and time variations in the medium are slow enough (in such a way that, locally both in space and in time, the medium can still be considered as approximately uniform and constant), we can replace the wave electric field (2.1) by a similar expression:

$$\vec{E}(\vec{r}, t) = \vec{E}_0(\vec{r}, t) \exp i\psi(\vec{r}, t) \tag{2.5}$$

where $\psi(\vec{r}, t)$ is the wave phase and $\vec{E}_0(\vec{r}, t)$ a slowly varying wave amplitude.

We can now define a local value for the wave frequency and for the wavevector, by taking the space and time derivatives of the phase function ψ:

$$\vec{k} = \frac{\partial}{\partial \vec{r}} \psi, \quad \omega = -\frac{\partial}{\partial t} \psi. \tag{2.6}$$

This local frequency ω and wavevector $\vec{k}$ are still related by a dispersion relation, which is now only locally valid:

$$R(\omega, \vec{k}; \vec{r}, t) = 0. \tag{2.7}$$

The parameters of the medium, for instance its refractive index or, in a plasma, its electron plasma frequency ω_p, depend on the position $\vec{r}$ and on time t. Such a dispersion relation is satisfied at every position and at each time. This means that its solution can be written as $\omega = \omega(\vec{r}, \vec{k}, t)$.

Starting from Maxwell's equations, it can be shown that this expression stays valid as long as the space and timescales for the variations of the medium are much slower than k^{-1} and ω^{-1}, or, in more precise terms, if the following inequality is satisfied:

$$\frac{2\pi}{\omega}\left|\frac{\partial}{\partial t}\ln\xi\right| + \frac{2\pi}{k}|\nabla\ln\xi| \ll 1 \tag{2.8}$$

where ξ is any scalar characterizing the background medium, for instance the refractive index or, for a plasma, the electron density.

It can easily be seen from the above definitions of the local frequency and wavevector that

$$\frac{\partial\vec{k}}{\partial t} = -\nabla\omega = -\left[\frac{\partial\omega}{\partial\vec{r}} + \frac{\partial\vec{k}}{\partial\vec{r}}\cdot\frac{\partial\omega}{\partial\vec{k}}\right]. \tag{2.9}$$

The last term in this equation was established by noting that, because ω is assumed to be a function of $\vec{r}$ and $\vec{k}$, it varies in space not only because of its explicit dependence on $\vec{r}$ but because the value of $\vec{k}$ is also varying. Let us introduce the definition of group velocity

$$\vec{v}_g = \frac{\partial\omega}{\partial\vec{k}}. \tag{2.10}$$

This is the velocity of the centroid of an electromagnetic wavepacket moving in the medium, $\vec{v}_g = d\vec{r}/dt$. The above equation can be written as

$$\left(\frac{\partial}{\partial t} + \vec{v}_g\cdot\frac{\partial}{\partial\vec{r}}\right)\vec{k} = -\frac{\partial\omega}{\partial\vec{r}}. \tag{2.11}$$

The differential operator on the left-hand side is nothing but the total time derivative d/dt. It means that we can rewrite the last two equations as

$$\frac{d\vec{r}}{dt} = \frac{\partial\omega}{\partial\vec{k}}, \qquad \frac{d\vec{k}}{dt} = -\frac{\partial\omega}{\partial\vec{r}}. \tag{2.12}$$

These equations of motion can be seen as describing the evolution of point particles: the photons. We notice that they are written in Hamiltonian form. The canonical variables are here the photon position $\vec{r}$ and the wavevector $\vec{k}$, while the frequency plays the role of the Hamiltonian function, $\omega \equiv h(\vec{r}, \vec{k}, t)$. In general, it will be time dependent according to

$$\frac{d\omega}{dt} = \frac{\partial \omega}{\partial t}. \tag{2.13}$$

These equations are well known in the literature and their derivation is given in textbooks [55] and in papers [8, 9, 117]. We see that they explicitly predict a frequency shift as well as a wavevector change. Equations (2.12) and (2.13) have, in fact, an obvious symmetrical structure. But, for some historical reason, the physical implications of such a structure for media varying both in space and time have only recently been fully understood [66].

These ray equations can also be written with implicit differentiation as

$$\frac{d\vec{k}}{dt} = \frac{\partial R/\partial \vec{r}}{\partial R/\partial \omega}, \quad \frac{d\vec{r}}{dt} = -\frac{\partial R/\partial \vec{k}}{\partial R/\partial \omega} \tag{2.14}$$

and

$$\frac{d\omega}{dt} = -\frac{\partial R/\partial t}{\partial R/\partial \omega}. \tag{2.15}$$

This implicit version, even if it is appropriate for numerical computations (this is currently used for magnetized plasmas with complicated spatial configurations), loses the clarity and elegance of the Hamiltonian approach. For this reason we will retain our attention on the explicit Hamiltonian version of the ray or photon equations.

2.2 Space and time refraction

In order to understand the physical meaning of the photon equations stated above, let us use a simple and illustrative example of a photon propagating in a non-dispersive (but space- and/or time-dependent) dielectric medium, described by the dispersion relation

$$\omega \equiv \omega(\vec{r}, \vec{k}, t) = \frac{kc}{n(\vec{r}, t)}. \tag{2.16}$$

From equations (2.12), we have

$$\frac{d\vec{r}}{dt} = \frac{\omega}{k^2}\vec{k}, \quad \frac{d\vec{k}}{dt} = \omega\frac{\partial}{\partial \vec{r}}\ln n. \tag{2.17}$$

From equation (2.13), we can also obtain

$$\frac{d\omega}{dt} = -\omega\frac{\partial}{\partial t}\ln n. \tag{2.18}$$

Let us successively apply these equations to three distinct situations: (i) an inhomogeneous but stationary medium; (ii) an homogeneous but time-dependent medium; (iii) an inhomogeneous and non-stationary medium.

2.2.1 Refraction

In the first situation, we assume two uniform and stationary media, with refractive indices n_1 and n_2, separated by a boundary layer of width l_b located around the plane $y = 0$. In order to describe such a configuration we can write

$$n(\vec{r}, t) \equiv n(y) = n_1 + \frac{\Delta n}{2}[1 + \tanh(k_b y)] \tag{2.19}$$

where $\Delta n = n_2 - n_1$, and $k_b = 2\pi/l_b$.

The hyperbolic tangent is chosen as a simple and plausible model for a smooth transition between the two media. Clearly, the ray equations (2.17, 2.18) are only valid when l_b is much larger than the local wavelength $2\pi/k$. On the other hand, if we are interested in the study of the photon or ray propagation across a large region with dimensions much larger than l_b, such a smooth transition can be viewed from far away as a sharp boundary, and the above model given by equation (2.19) corresponds to the usual optical configuration for wave refraction.

In this large scale view of refraction, the above law can be approximated by

$$n(y) = n_1 + \Delta n H(y) \tag{2.20}$$

where $H(y) = 0$ for $y < 0$ and $H(y) = 1$ for $y > 0$ is the well-known step function or Heaviside function.

First of all, it should be noticed that from equations (2.18) and (2.19) we have

$$\frac{d\omega}{dt} = 0. \tag{2.21}$$

This means that the wave frequency is a constant of motion $\omega(\vec{r}, \vec{k}, t) = \omega_0 = $ const. If the plane of incidence coincides with $z = 0$, the time variation of the wavevector components is determined by

$$\frac{dk_x}{dt} = 0 \tag{2.22}$$

$$\frac{dk_y}{dt} = \omega_0 \frac{\partial}{\partial y} \ln n(y) = \frac{\omega_0 \Delta n}{2n(y)} k_b \operatorname{sech}^2(k_b y). \tag{2.23}$$

We see that the wavevector component parallel to the gradient of the refractive index is changing across the boundary layer, and that the perpendicular component remains constant. Defining $\theta(y)$ as the angle between the wavevector $\vec{k}$ and the normal to the boundary layer $\hat{e}_y$, we can obtain, from the first of these equations,

$$k_x = \frac{\omega_0}{c} n(y) \sin\theta(y) = \text{const.} \tag{2.24}$$

Considering the asymptotic values of $\theta(y)$ as the usual angles of incidence and of transmission $\theta_1 = \theta(y \to -\infty)$ and $\theta_2 = \theta(y \to +\infty)$, we reduce this result to the well-known Snell's law of refraction

$$n_1 \sin\theta_1 = n_2 \sin\theta_2. \tag{2.25}$$

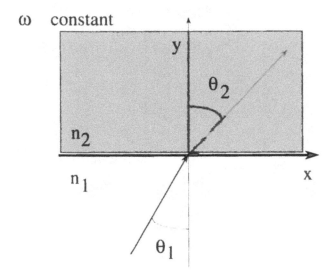

ω constant

Figure 2.1. Photon refraction at a boundary between two stationary media.

Returning to equations (2.17) we can also see that refraction leads to a change in the group velocity, or photon velocity, with asymptotic values $v_g = c/n_1$, for $y \rightarrow -\infty$, and $v_g = c/n_2$, for $y \rightarrow +\infty$. In a broad sense, we could be led to talk about photon acceleration during refraction. But, as we will see later, this is not appropriate because this change in group velocity is exactly compensated by a change in the photon effective mass, in such a way that the total photon energy remains constant. This will not be the case for the next two examples.

2.2.2 Time refraction

Let us turn to the opposite case of a medium which is uniform in space but changes its refractive index with time. We can describe this change by a law similar to equation (2.19):

$$n(\vec{r}, t) \equiv n(t) = n_1 + \frac{\Delta n}{2} [1 + \tanh(\Omega_b t)]. \tag{2.26}$$

Here, the timescale for refractive index variation $2\pi/\Omega_b$ is much larger than the wave period $2\pi/\omega$. From equation (2.19), we now have

$$\frac{d\omega}{dt} = -\omega \frac{\partial}{\partial t} \ln n(t) = -\frac{\omega \Delta n}{2n(t)} \Omega_b \operatorname{sech}^2(\Omega_b t). \tag{2.27}$$

We see that the photon frequency is shifted as time evolves, following a law similar to that of the wavevector change during refraction. On the other hand, if

k = constant

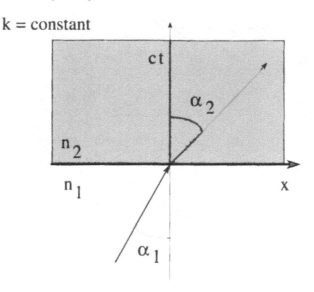

Figure 2.2. Time refraction of a photon at the time boundary between two uniform media.

propagation is taken along the x-direction, we can see from equation (2.18) that

$$\frac{dx}{dt} = \frac{c}{n(t)} = v_g(t), \qquad \frac{dk}{dt} = 0. \tag{2.28}$$

The photon momentum is now conserved and the change in the group velocity is only related to the frequency (or energy) shift. Snell's law (2.24) is now replaced by

$$k = \frac{\omega(t)}{c} n(t) = \text{const.} \tag{2.29}$$

Defining the initial and the final values for the frequency as the asymptotic values $\omega_1 = \omega(t \to -\infty)$ and $\omega_2 = \omega(t \to +\infty)$, we can then write

$$n_1 \omega_1 = n_2 \omega_2. \tag{2.30}$$

This can be called the *Snell's law for time refraction*. Actually, in the plane (x, ct) we can define an angle α, similar to the usual angle of incidence θ defined in the plane (x, y), such that $\tan \alpha = v_g/c = 1/n$. This could be called the angle of temporal incidence. Replacing it in equation (2.30), we get

$$\omega_1 \tan \alpha_2 = \omega_2 \tan \alpha_1. \tag{2.31}$$

This equation resembles the well-known Snell's law of refraction, equation (2.25). However, there is an important qualitative difference, related to the mechanism of total reflection. We know that this can occur when a photon propagates in a medium of decreasing refractive index. From equation (2.25) we see that, for $n_1 > n_2$, it is possible to define a critical angle θ_c such that $\sin\theta_c = n_2/n_1$. For $\theta_1 \geq \theta_c$ we would have $\sin\theta_2 \geq 1$ which means that propagation is not allowed in medium 2 and that total reflection at the boundary layer will occur.

In contrast, equations (2.30) or (2.31) are always satisfied for arbitrary (positive) values of the asymptotic refractive indices n_1 and n_2. Not surprisingly, no such thing as a reflection back in time (or a return to the past) can ever occur. This is the physical meaning of the replacement of the sine law for reflection by the tangent law for time refraction.

2.2.3 Space–time refraction

Let us now assume a more general situation where the optical medium is both inhomogeneous in space and non-stationary in time. As a simple generalization of the previous two models, we will assume that the boundary layer between media 1 and 2 is moving with a constant velocity u along the x-direction.

Equations (2.19) and (2.26) are now replaced by a similar expression, of the form

$$n(\vec{r}, t) \equiv n(x - ut) = n_1 + \frac{\Delta n}{2}[1 + \tanh(k_b x - \Omega_b t)] \qquad (2.32)$$

with $\Omega_b = k_b u$. From the ray equations (2.12) we can now derive a simple relation between the time variation of frequency and wavenumber:

$$\frac{dk}{dt} = \frac{1}{u}\frac{d\omega}{dt}. \qquad (2.33)$$

This relation means that, when a photon interacts with a moving boundary layer, the wave frequency and wavevector are both shifted, in a kind of space–time refraction. We notice that in order to establish the Snell's laws (2.24, 2.25), or their temporal counterparts (2.29, 2.30), we had to define some constant of motion.

In the first case, the constants of motion were the photon frequency ω and the wavevector component parallel to the boundary layer k_x. In the second case, the constant of motion was the photon wavevector. Now, for the case of refraction in a moving boundary, or space–time refraction, we need to find a new constant of motion because neither the frequency nor the wavevector are conserved.

This new constant of motion can be directly obtained from equation (2.33), but it is useful to explore the Hamiltonian properties of the ray equations (2.12, 2.13). For such a purpose, let us then define a canonical transformation from the variables (x, k) to the new pair of variables (x', k'), such that

$$x' = x - ut, \quad k' = k. \qquad (2.34)$$

The resulting canonical ray equations can be written as

$$\frac{dx'}{dt} = \frac{\partial \omega'}{\partial k'}, \quad \frac{dk'}{dt} = -\frac{\partial \omega'}{\partial x'}.$$ (2.35)

It can easily be seen that the new Hamiltonian ω' appearing in these equations is a time-independent function defined by

$$\omega' \equiv \omega'(x', k') = \omega(x', k') - uk' = \left[\frac{c}{n(x')} - u \right] k'.$$ (2.36)

In a more formal way, we can define a generating function for this canonical transformation, as

$$F(x, k', t) = (x - ut)k'$$ (2.37)

such that

$$k = \frac{\partial F}{\partial x} = k'$$ (2.38)

$$x' = \frac{\partial F}{\partial k'} = x - ut.$$ (2.39)

The Hamiltonian (2.36) is determined by

$$\omega' = \omega + \frac{\partial F}{\partial t} = \omega - uk'.$$ (2.40)

Because, based on this canonical transformation, we were able to define a new constant of motion $\omega' = $ const, we can take the asymptotic values of $n(x')$ and k' for regions far away from the boundary layer ($x' \to \pm\infty$), and write

$$\omega' = \left(\frac{c}{n_1} - u \right) k_1 = \left(\frac{c}{n_2} - u \right) k_2$$ (2.41)

or, equivalently, in terms of the frequency

$$\omega' = \omega_1 \left(1 - \frac{n_1}{c} u \right) = \omega_2 \left(1 - \frac{n_2}{c} u \right).$$ (2.42)

We are then led to the following formula for the total frequency shift associated with space–time refraction:

$$\omega_2 = \omega_1 \frac{1 - \beta_1}{1 - \beta_2}$$ (2.43)

where $\beta_i = un_i/c$, for $i = 1, 2$. We see that the resulting frequency up-shift can be extremely large if the boundary layer and the photons are moving in the same direction at nearly the same velocity: $\beta_2 \simeq 1$.

When compared with the frequency shifts associated with the case of a purely time refraction, this shows important qualitative and quantitative differences. This new effect of almost unlimited frequency shift is a clear consequence of the combined influence of the space and time variations of the medium, or in other words, of the synergy between the usual refraction and the new time refraction considered above.

A comment has to be made on the exact resonance condition, defined by $\beta_2 = 1$. At a first impression, it could be concluded from equation (2.43) that the frequency shift would become infinitely large if this resonance condition were satisfied. However, because in this case the photons are moving with the same velocity as the boundary layer, they would need an infinite time to travel across the layer and to be infinitely frequency shifted.

A closer look at the resonant photon motion shows that this trajectory would be $x' = $ const and, according to equation (2.36), no frequency shift would occur. On the other hand, the resonant photon trajectory is physically irrelevant because it represents an ensemble of zero measure in the photon phase space (x, k). What is relevant is that, in the close vicinity of this particular trajectory, a large number of possible photon trajectories will lead to extremely high-frequency transformations taking place in a finite but large amount of time.

Let us now apply the above analysis to the well-known and famous problem of photon reflection by a relativistic mirror. This can be done by assuming that $n_1 = 1$ and Δn is such that $n(x') = 0$ for $x' \geq 0$.

The mirror will move with constant velocity $-u$ and its surface will coincide with the plane $x' = 0$. A photon propagating initially in a vacuum in the opposite direction, with a wavevector $\vec{k} = k_1 \hat{e}_x$, will be reflected by the mirror and come back to medium 1 with a wavevector $\vec{k} = -k_1 \hat{e}_x$.

This means that in the above equation (2.43) we can make $\beta_2 = \beta = u/c$ and $\beta_1 = -\beta$. It will then become

$$\omega_2 = \omega_1 \frac{1 + \beta}{1 - \beta}. \tag{2.44}$$

This is nothing but the well-known formula for the relativistic mirror, which was derived here in a very simple way, without invoking relativity or using any Lorentz transformation. Such a result is possible because the photon equations are exactly relativistic by nature. Apart from providing a very simple and alternative way to derive the relativistic mirror effect, this result is also interesting because it shows that such an effect can be seen as a particular case of the more general space–time refraction or photon acceleration processes.

2.3 Generalized Snell's law

We generalize here the above discussion of space–time refraction, by considering oblique photon propagation with respect to the moving boundary between two different dielectric media. The moving boundary will now be described by the following refractive index:

$$n(\vec{r}, t) = n_1 + \frac{\Delta n}{2}\left[1 + \tanh(\vec{k}_b \cdot \vec{r} - \Omega_b t)\right] \tag{2.45}$$

where $\Omega_b = \vec{k}_b \cdot \vec{u}$ and $\vec{u}$ is the velocity of the moving boundary. As before, we assume that the velocity and the slope of the boundary layer, defined by $\vec{k}_b$, remain constant.

It is useful to introduce parallel and perpendicular propagation, by defining $\vec{r} = r_\parallel \hat{b} + \vec{r}_\perp$ and $\vec{k} = k_\parallel \hat{b} + \vec{k}_\perp$, where $\hat{b} = \vec{k}_b / k_b$. We conclude from the photon equations of motion that

$$\frac{dk_\parallel}{dt} = \frac{kc}{n^2}\frac{\partial n}{\partial r_\parallel} \tag{2.46}$$

$$\frac{d\vec{k}_\perp}{dt} = 0. \tag{2.47}$$

The conservation of the perpendicular wavevector is due to the fact that the refractive index is only a function of $r_\parallel$: $n(\vec{r}, t) = n(r_\parallel, t)$. We can now use the canonical transformation from $(\vec{r}, \vec{k})$ to a new pair of variables $(\vec{r}', \vec{k}')$, generated by the transformation function

$$F(\vec{r}, \vec{k}', t) = (\vec{r} - \vec{u}t) \cdot \vec{k}'. \tag{2.48}$$

This is a straightforward generalization of the canonical transformation used for the one-dimensional problem. This leads to

$$\vec{k} = \frac{\partial F}{\partial \vec{r}} = \vec{k}' \tag{2.49}$$

$$\vec{r}' = \frac{\partial F}{\partial \vec{k}'} = \vec{r} - \vec{u}t. \tag{2.50}$$

The new Hamiltonian $\omega'(\vec{r}', \vec{k}', t)$ appearing in these equations is determined by

$$\omega' = \omega + \frac{\partial F}{\partial t} = \omega - \vec{u} \cdot \vec{k}'. \tag{2.51}$$

This Hamiltonian is a constant of motion because in the new coordinates neither n nor ω depend explicitly on time. It is clear that we can now define two constants of motion for photons crossing the moving boundary. Let us call them I_1 and I_2. The first one is simply the new Hamiltonian ω', and the other one is the perpendicular wavenumber $k_\perp$:

$$I_1 = \omega(\vec{r}', \vec{k}') - \vec{u} \cdot \vec{k}' \tag{2.52}$$

$$I_2 = |\vec{k}' \times \hat{b}|. \tag{2.53}$$

Notice that our analysis is valid even if the velocity of the moving boundary $\vec{u}$ is not parallel to the gradient of the refractive index $\hat{b}$. However, in order to place our discussion on simple grounds, we assume here that $\vec{u}$ is parallel to $\hat{b}$. If $\theta(\vec{r}')$ is the angle between the photon wavevector and the front velocity, defined for each photon position $\vec{r}'$ along the trajectory, we can rewrite the two invariants as

$$I_1 = \omega(\vec{r}'\ \vec{k}')\left[1 - \frac{u}{c}n(\vec{r}')\cos\theta(\vec{r}')\right] \tag{2.54}$$

$$I_2 = k'\sin\theta(\vec{r}'). \tag{2.55}$$

In the first of these equations, we made use of the local dispersion relation $n(\vec{r}')\omega(\vec{r}', \vec{k}') = k'c$. Let us retain our attention on photon trajectories which start in medium 1 far away from the boundary, at $\vec{r}' \rightarrow -\infty$, cross the boundary layer (at $\vec{r}' = 0$) and then penetrate in medium 2 moving towards $\vec{r}' \rightarrow +\infty$.

Denoting by the subscripts 1 and 2 the frequencies and angles at the extreme ends of such trajectories, we can conclude, from the above two invariants, that

$$\omega_2 = \omega_1 \frac{1 - \beta_1 \cos\theta_1}{1 - \beta_2 \cos\theta_2} \tag{2.56}$$

where $\beta_i = (u/c)n_i$, for $i = 1, 2$. This expression establishes the total frequency shift. We can also conclude that

$$n_1 \sin\theta_1 = \frac{\omega_2}{\omega_1}n_2 \sin\theta_2. \tag{2.57}$$

This last equation can be seen as the generalized Snell's law, valid for refraction at a moving boundary between two different dielectric media. It shows that the relation between the angles of incidence and transmission depends not only on the refractive indices of the two media but also on the photon frequency shift.

Of course, when the velocity of the boundary layer tends to zero, $u \rightarrow 0$, the frequency shift also tends to zero, $\omega_2 \rightarrow \omega_1$, as shown by equation (2.56). In this case equation (2.57) reduces to the usual form of Snell's law (2.25), as it should. On the other hand, for a finite velocity $u \neq 0$, but for normal incidence $\theta_1 = \theta_2 = 0$, equation (2.56) reduces to $\omega_2 = \omega_1(1 - \beta_1)/(1 - \beta_2)$, in accordance with the discusssion of the previous section.

We should notice that equation (2.56) is valid for an arbitrary value of the velocity of the moving boundary. It stays valid, in particular, for supraluminous moving boundaries, such that $u > c$. It is well known that such boundaries can exist without violating the principles of Einstein's theory of relativity [31]. They can even be built in a medium where the atoms are completely at rest, as will be discussed in the next chapter.

Let us consider the limit of an infinite velocity $u \rightarrow \infty$. Equation (2.56) is then reduced to equation (2.30): $n_2\omega_2 = n_1\omega_1$. This means that it reproduces the above result for a pure time refraction.

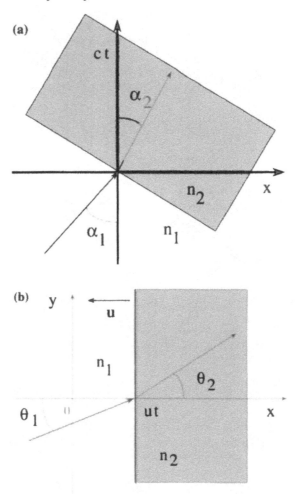

Figure 2.3. Illustration of the generalized Snell's law: photon refraction at the moving boundary between two uniform media. (a) Diagram (x, ct) for one-dimensional propagation; (b) oblique propagation.

Now, for an arbitrary oblique incidence, but for highly supraluminous fronts, $u \gg c$, we get from equations (2.56, 2.57)

$$\sin(\theta_1 - \theta_2) \simeq 0. \tag{2.58}$$

We see that, in this limit of a highly supraluminous front, there is no momentum transfer from the medium to the photon, $\theta_1 \simeq \theta_2$. The resulting frequency shift is determined in this case by the law of pure time refraction, showing that

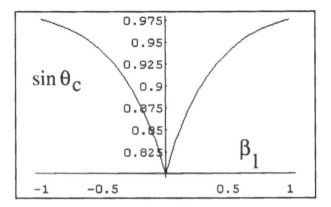

Figure 2.4. Dependence of the critical angle θ_c on the front velocity β_1, for $n_2/n_1 = 0.8$.

space changes of the refractive index of the medium do not contribute to the final frequency shift. The dominant processes are associated with the time changes.

Let us now examine the conditions for the occurrence of total reflection. From the above expression of the generalized Snell's law (2.57) we can define a critical angle $\theta_1 = \theta_c$ such that $\theta_2 = \pi/2$, or

$$\sin \theta_c = \frac{n_2}{n_1} \frac{\omega_2}{\omega_1}. \tag{2.59}$$

Using equation (2.56), we also see that

$$\sin \theta_c = \frac{n_2}{n_1}(1 - \beta_1 \cos \theta_c). \tag{2.60}$$

This can be explicitly solved and gives

$$\sin \theta_c = \frac{n_1 n_2}{n_1^2 + \beta_1^2 n_2^2} \left\{ 1 + |\beta_1| \sqrt{1 - \left(\frac{n_2}{n_1}\right)^2 (1 - \beta_1^2)} \right\}. \tag{2.61}$$

For a boundary at rest, $\beta_1 = 0$, this expression reduces to the well-known formula for the critical angle $\sin \theta_c = n_2/n_1$. We also notice that this result guarantees that in the absence of any boundary (which means $n_2 = n_1$) we always get $\sin \theta_c = 1$, or $\theta_c = \pi/2$ (absence of total reflection) for arbitrary values of β_1, as expected. The same occurs for purely time-varying media ($|\beta_i| \to \infty$), confirming that no time reflection is possible.

For angles of incidence larger than this critical angle $\theta_1 > \theta_c$, the photon will be reflected. This means that the direction of the reflected wavevector will be reversed with respect to that assumed in equation (2.56).

If the boundary layer is moving towards the incident photon (which is the usual configuration of the relativistic mirror effect), we have to replace β_1 by $-\beta_1$

in this equation. Moreover, we will replace the subscript 2 by the subscript 3, in order to stress that the final photon state is given by reflection and not by transmission:

$$\omega_3 = \omega_1 \frac{1 + \beta_1 \cos\theta_1}{1 - \beta_1 \cos\theta_3}. \tag{2.62}$$

Here the angle of reflection θ_3 is not equal to the angle of incidence (as occurs for a stationary boundary) but, according to equation (2.57), it is given by

$$\sin\theta_3 = \frac{\omega_1}{\omega_3} \sin\theta_1. \tag{2.63}$$

These two formulae state the relativistic mirror effect for oblique incidence, as a particular case of photon acceleration at the moving boundary. Actually, these formulae stay valid even for partial reflection at the boundary, for angles of incidence smaller than the critical angle. But in order to account for partial reflection we will have to use a full wave description and not use the basis of the geometric optics approximation.

Let us return to the general expression for the two photon invariants I_1 and I_2, and make a short comment on the case where the vector velocity, $\vec{u}$, and $\vec{k}_b$, characterizing the gradient of the refractive index across the boundary layer, are not parallel to each other. We can then write $\vec{u} \cdot \vec{k}' = uk' \cos\alpha(\vec{r}')$ and $|\vec{k}' \times \hat{b}| = k' \sin\theta(\vec{r}')$, where these two angles $\alpha(\vec{r}')$ and $\theta(\vec{r}')$ are not necessarily equal to each other. If $\vec{u}$ is contained in the plane of incidence, we have $\theta(\vec{r}') = \alpha(\vec{r}') + \theta_0$, where θ_0 is a constant. In the expressions of the two invariants I_1 and I_2, as given by equation (2.55), the angle $\theta(\vec{r}')$ has to be replaced by $\alpha(\vec{r}') = \theta(\vec{r}') - \theta_0$.

In our discussion we have kept our attention on very simple but paradigmatic situations of boundary layers moving with constant velocity. This is because our main objective here was to extend the existing geometric optics theory to the space–time domain and to obtain simple but also surprising generalizations of well-known formulae and well-known concepts.

Our Hamiltonian approach to photon dynamics can however be applied to more complicated situations of space–time varying media, for instance to non-stationary and accelerated boundary layers.

2.4   Photon effective mass

Let us now go deeper in the analysis of the photon dynamics and explore the analogies between the photon ray equations and the equations of motion of an arbitrary point particle with finite rest mass. From the above definitions of the local wavevector $\vec{k}$ and local frequency ω, it is obvious that the total variation of the phase ψ, as a function of the space and time coordinates, is

$$d\psi = \vec{k} \cdot d\vec{r} - \omega \, dt. \tag{2.64}$$

If we identify the photon frequency ω with the particle Hamiltonian, and the wavevector $\vec{k}$ with its momentum, as we did before, we can easily see from classical mechanics [33, 56] that this phase is nothing but the photon action

$$\psi = \int \vec{k} \cdot d\vec{r} - \omega(\vec{r}, \vec{k}, t) \, dt. \tag{2.65}$$

In order to establish the photon trajectories we can apply the principle of minimum action, and state that the phase has to obey the extremum condition

$$\delta\psi = 0. \tag{2.66}$$

It is well known that, from this variational principle, it is possible to rederive the photon canonical equations already established in section 2.1. On the other hand, for a dielectric medium with refractive index $n = kc/\omega$, and from the definitions of $\vec{k}$ and ω, we can get the following equation for the phase:

$$\left(\frac{\partial\psi}{\partial\vec{r}}\right)^2 = \frac{n^2}{c^2}\left(\frac{\partial\psi}{\partial t}\right)^2. \tag{2.67}$$

In the particular case of a stationary medium, we have $\omega = \omega_0 = $ const, and equation (2.65) reduces to

$$\psi = \psi_0 - \omega_0 t = \int \vec{k} \cdot d\vec{r} - \omega_0 t \tag{2.68}$$

where ψ_0 is the reduced photon action. Equation (2.67) is then reduced to the usual eikonal equation of geometric optics [15]

$$\left(\frac{\partial\psi_0}{\partial\vec{r}}\right)^2 = \frac{n^2}{c^2}\omega_0^2. \tag{2.69}$$

We can then refer to the more general equation for the phase, equation (2.67), as for the generalized eikonal equation, valid for space–time varying media.

We also know from classical mechanics that the action can be defined as the time integral of a Lagrangian function. The photon Lagrangian is then determined by

$$\psi = \int_{t_1}^{t_2} L(\vec{r}, \vec{v}, t) \, dt \tag{2.70}$$

where $\vec{v} = d\vec{r}/dt$ is the photon velocity, or group velocity. Comparing this with equation (2.65), we obtain

$$L(\vec{r}, \vec{v}, t) = \vec{k} \cdot \vec{v} - \omega(\vec{r}, \vec{k}, t). \tag{2.71}$$

It should be noticed that, in this expression, the photon wavevector is not an independent variable because it is determined by

$$\vec{k} = \frac{\partial L}{\partial\vec{v}}. \tag{2.72}$$

We can also determine the force acting on the photon as the derivative of the Lagrangian with respect to the coordinates (or the gradient of L):

$$\vec{f} = \frac{\partial L}{\partial \vec{r}}. \tag{2.73}$$

With these definitions we can reduce the Lagrangian equation

$$\frac{d}{dt}\frac{\partial L}{\partial \vec{v}} - \frac{\partial L}{\partial \vec{r}} = 0 \tag{2.74}$$

to Newton's equation of motion

$$\frac{d\vec{k}}{dt} = \vec{f} = -\frac{\partial \omega}{\partial \vec{r}}. \tag{2.75}$$

Obviously, this also agrees with the second of the Hamiltonian ray equations (the first one can be seen as the definition of $\vec{v} = d\vec{r}/dt$). We have given here the three equivalent versions of the photon equations of motion in a space-time varying medium: (1) the Newtonian equation (2.75); (2) the Lagrangian equation (2.74) and (3) the Hamiltonian equations already stated in section 2.1. This means that the photon trajectories can be described by the formalism of classical mechanics, in a way very similar to that of point particles with finite mass.

Let us explore further this analogy between the photon and a classical particle. The example of high-frequency transverse photons in an isotropic plasma is particularly interesting because of the simple form of the associated dispersion relation: $\omega = \sqrt{k^2 c^2 + \omega_p^2(\vec{r}, t)}$. If we multiply this expression by Planck's constant (divided by 2π), redefine the photon energy as $\epsilon = \hbar\omega$ and the photon momentum as $\vec{p} = \hbar\vec{k}$, we get

$$\epsilon = \sqrt{p^2 c^2 + \hbar^2 \omega_p^2(\vec{r}, t)} = \sqrt{p^2 c^2 + m_{\text{eff}}^2 c^4}. \tag{2.76}$$

This means that the photon in a plasma is a relativistic particle with an effective mass defined by

$$m_{\text{eff}} = \omega_p \hbar / c^2. \tag{2.77}$$

We see that the photon mass in a plasma is proportional to the plasma frequency (or to the square root of the electron plasma density). Therefore, in a non-stationary and non-uniform medium, this mass is not a constant because it depends on the local plasma properties.

In the following, we will use $\hbar = 1$. We can also see that the photon velocity in this medium is determined by $v = pc^2/\epsilon$, or $v = kc^2/\omega$. The corresponding relativistic gamma factor is then

$$\gamma = \left(1 - \frac{1}{c^2}\left|\frac{d\vec{r}}{dt}\right|^2\right)^{-1/2} = \left(1 - \frac{k^2 c^2}{\omega^2}\right)^{-1/2} = \frac{\omega}{\omega_p}. \tag{2.78}$$

We can also adapt Einstein's famous formula for the energy of a relativistic particle to the case of a photon moving in a plasma, as

$$\omega = \gamma \omega_p = m_{eff} \gamma c^2. \tag{2.79}$$

Let us now write the photon Lagrangian in an explicit form. Using equation (2.71), we have

$$L = \frac{v}{c}\sqrt{\omega^2 - \omega_p^2} - \omega = -\omega_p\left(\gamma - \frac{v}{c}\sqrt{\gamma^2 - 1}\right). \tag{2.80}$$

This can also be written as

$$L(\vec{r}, \vec{v}, t) = -\frac{1}{\gamma}m_{eff}c^2 = -\omega_p(\vec{r}, t)\left(1 - \frac{v^2}{c^2}\right)^{1/2}. \tag{2.81}$$

The photon momentum $\vec{k}$, and the force acting on the photon $\vec{f}$, are determined by

$$\vec{k} = \frac{\partial L}{\partial \vec{v}} = -\omega_p\frac{\partial}{\partial \vec{v}}\left(1 - \frac{v^2}{c^2}\right)^{1/2} = \frac{\omega_p}{c^2}\gamma\vec{v} = m_{eff}\gamma\vec{v} \tag{2.82}$$

and

$$\vec{f} = \frac{\partial L}{\partial \vec{r}} = -\left(1 - \frac{v^2}{c^2}\right)^{1/2}\frac{\partial \omega_p}{\partial \vec{r}} = -\frac{\omega_p}{\omega}\frac{\partial \omega_p}{\partial \vec{r}}. \tag{2.83}$$

The Newtonian equation of motion (2.75) can then be written as

$$\frac{d\vec{k}}{dt} = -\frac{\omega_p}{\omega}\frac{\partial \omega_p}{\partial \vec{r}}. \tag{2.84}$$

The same kind of description can be used for electromagnetic wave propagation in a waveguide. The effective mass of the confined photons will then be equal to $m_{eff} = \omega_0/c^2$, where ω_0 is the cut-off frequency for the specific mode of propagation.

The same equations of motion apply here, but only for one-dimensional propagation (along the waveguide structure) and, for non-stationary and non-uniform waveguides, the same effective mass changes and frequency shifts are expected. The implications of the effective mass of a photon in a waveguide were considered, in a somewhat speculative maner, by Rivlin [90].

In general terms, a photon always propagates in a medium with velocity less than c. In the language of modern quantum field theory it can be considered as a 'dressed' photon, because of the polarization cloud. For a generic dispersive dielectric medium, the photon velocity is determined by

$$\vec{v} \equiv \frac{\partial \omega}{\partial \vec{k}} = \frac{\partial}{\partial \vec{k}}\frac{kc}{n} = \frac{\partial}{\partial \vec{k}}\frac{kc}{\sqrt{1 + \chi(\omega)}} \tag{2.85}$$

where $\chi(\omega)$ is the susceptibility of the medium. We then have

$$v = \frac{c\sqrt{1+\chi}}{(1+\chi)+(\omega/2)(\partial\chi/\partial\omega)}. \tag{2.86}$$

The relativistic γ factor associated with the photon in the medium is then, by definition,

$$\gamma = \left(1 - \frac{v^2}{c^2}\right)^{-1/2} = \frac{(1+\chi)+(\omega/2)(\partial\chi/\partial\omega)}{\{[(1+\chi)+(\omega/2)(\partial\chi/\partial\omega)]^2 - (1+\chi)\}^{1/2}}. \tag{2.87}$$

In the case of a plasma, we have $\chi = -\omega_p^2/\omega^2 = -1/\gamma^2$. This is in agreement with equation (2.78), as expected. For a non-dispersive medium we have $\partial\chi/\partial\omega = 0$, and we get the following simple result:

$$\gamma = \left(1 + \frac{1}{\chi}\right)^{1/2} = \frac{n}{\sqrt{n^2-1}}. \tag{2.88}$$

The gamma factor tends to infinity when the refractive index tends to one (the case of a vacuum) because the photon velocity then becomes equal to c. Let us now establish a definition of the photon effective mass, valid for a dispersive optical medium. Using Einstein's energy relation for a relativistic particle, we get

$$m_{\text{eff}} = \frac{\omega}{\gamma c^2}. \tag{2.89}$$

This means that, in general, m_{eff} is a function of the frequency ω. Exceptions are the isotropic plasma and the waveguide cases. As a consequence, we are not allowed to write the dispersion relation as $\omega = (k^2c^2 + m_{\text{eff}}^2 c^4)^{1/2}$, except for these important but still exceptional cases, because for a refractive index larger than one this would simply imply imaginary effective masses, in contrast with the well-behaved definition stated above. For instance, for a non-dispersive dielectric medium with $n > 1$ we have, from equation (2.89),

$$m_{\text{eff}} = \frac{\omega}{nc^2}\sqrt{n^2-1}. \tag{2.90}$$

Let us now write the Lagrangian for a photon in a generic dielectric medium. Using equation (2.71), we have

$$L = -\omega\left(1 - \frac{v}{c}\sqrt{1+\chi}\right). \tag{2.91}$$

This reduces to equation (2.80) for a plasma, where $\chi = -\omega_p^2/\omega^2$. In contrast, for a non-dispersive medium, we have $v = c/\sqrt{1+\chi}$ and the Lagrangian is equal to zero. Notice that, in general, we are not allowed to write $L = -m_{\text{eff}}c^2/\gamma$, except for plasmas and waveguides.

From equation (2.73) we can calculate the force acting on the photon

$$\vec{f} = \frac{\partial L}{\partial \vec{r}} = \omega \frac{\partial}{\partial \vec{r}} \ln n = \frac{1}{2} \frac{\omega}{(1 + \chi)} \frac{\partial \chi}{\partial \vec{r}}. \tag{2.92}$$

This reduces to equation (2.83) for $\chi = -\omega_p^2/\omega^2$.

We will now make a final comment on the nature of this force. For a non-uniform but stationary medium, the force acting on the photon is responsible for the change in the photon velocity according to the usual laws of refraction: a gradient of the refractive index always leads to a change in the photon velocity, as stated by this expression for the force. We have then a variation in the relativistic γ factor: $\partial \gamma/\partial t \neq 0$.

However this is not equivalent to photon acceleration, because the total energy of the photon remains unchanged: $\partial \omega/\partial t = 0$. The reason is that the variation in velocity (or in kinetic energy) is exactly compensated by an equal and opposite variation in the effective mass (or in the rest energy)

$$\frac{\partial}{\partial t} \ln \gamma = -\frac{\partial}{\partial t} \ln m_{\text{eff}}. \tag{2.93}$$

If, in contrast, the properties of the medium are also varying with time, this equality is broken and we can say that photon acceleration takes place. The medium exchanges energy with the photon, and the photon frequency (or its total energy) is not conserved: $\partial \omega/\partial t \neq 0$.

2.5   Covariant formulation

The similarities between space and time boundaries noted in the preceding sections suggest and almost compel the use of four-vectors, defined in relativistic space–time. Let us then define the four-vector position and momentum as

$$x^i = \{ct, \vec{r}\}, \quad k^i = \left\{\frac{\omega}{c}, \vec{k}\right\}. \tag{2.94}$$

The flat space–time of special relativity [43, 55] is described by the Minkowski metric tensor g^{ij}, such that $g^{00} = -1$, $g^{ii} = 1$ (for $i = 1, 2, 3$) and $g^{ij} = 0$ (for $i \neq j$). From the relation $k^i = g^{ij}k_j$, we get $k^i = k_i$ (for $i = 1, 2, 3$) and $k^0 = -k_0 = \omega/c$.

The total variation of the photon phase (or action) will then be given by

$$d\psi = \vec{k} \cdot d\vec{r} - \omega \, dt = \sum_{i=1}^{3} k_i \, dx^i - k^0 \, dx^0 = k_i \, dx^i \tag{2.95}$$

where, in the last expression, we have used Einstein's summation rule for repeated indices.

We can then establish an expression for the variational principle (2.66), as

$$\delta\psi = \delta\int k_i\, dx^i = 0. \tag{2.96}$$

The photon canonical equations can now be written as

$$\frac{dx^i}{dx^0} = -\frac{\partial k_0}{\partial k_i}, \quad \frac{dk^i}{dx^0} = \frac{\partial k_0}{\partial x_i}. \tag{2.97}$$

Notice that, for $i = 0$, the second of these equations simply states that $d\omega/dt = \partial\omega/\partial t$, and the first one is an identity. For $i = 1, 2, 3$, we obtain the above three-dimensional canonical equations.

It should also be noticed that, in these equations, the momentum component k_0 is not an independent variable, but a function of the other variables $k_0 \equiv k_0(k_1, k_2, k_3, x^i)$. The explicit form of this function depends on the properties of the medium. For a generic dielectric medium, we have

$$k_0 = -\frac{1}{n(k_0, x^i)}\left(\sum_{j=1}^{3} k_j k^j\right)^{1/2}. \tag{2.98}$$

We can certainly recognize here the dispersion relation $\omega = kc/n$, written in a more sophisticated notation. The square of the four-vector momentum is determined by

$$k_i k^i = -(k_0)^2 + \sum_{j=1}^{3}(k_i)^2 = k^2 - \left(\frac{\omega}{c}\right)^2. \tag{2.99}$$

Noting that $k = \omega n/c$, and using $n = \sqrt{1 + \chi}$, we obtain

$$k_i k^i = (k_0)^2 \chi(k_0, x^i). \tag{2.100}$$

This is the dispersion relation in four-vector notation. In the case of a plasma, we simply have

$$k_i k^i = -\frac{\omega_p^2}{c^2} = -m_{\text{eff}}^2 c^2. \tag{2.101}$$

Now, the moving boundary between two dielectric media can be described by the law

$$n(x^i) = n_1 + \frac{\Delta n}{2}\left[1 + \tanh(K_j x^j)\right] \tag{2.102}$$

where $K^j = \{\Omega_b/c = \bar{k}_b \cdot \bar{u}/c, \bar{k}_b\}$ is the four-vector defining the space–time properties of that boundary.

The invariant I_1 can also be written in the same notation as

$$I_1 = -u_i k^i \tag{2.103}$$

where the four-vector velocity $u^i = \{c, \vec{u}\}$ was used.

What we have written until now is nothing but the equations already established for the photon equations of motion in the new language of the four-dimensional space–time of special relativity. This new way of writing can be useful in, at least, two different ways.

First, it stresses the fact that the apparently distinct phenomena of refraction and photon acceleration are nothing but two different aspects of a more general physical feature: the photon refraction due to a variation of the refractive index in the four-dimensional relativistic space–time. Second, it can be useful for future generalizations of photon equations to the case of a curved space–time, where gravitational effects can also be included.

We can now make a further qualitative jump. Given the formal analogy between equation (2.95) and the phase defined in the usual three-dimensional space coordinates, we can generalize it and write

$$d\psi = k_i \, dx^i - h(x^i, k_i) \, d\tau. \qquad (2.104)$$

Here we have used a timelike variable τ, to be identified later with the photon proper time, and a new Hamiltonian function such that $h(x^i, k_i) = 0$. This function is sometimes called a super-Hamiltonian [33, 76].

The covariant form of the variational principle (2.96) can now be written as

$$
\begin{aligned}
\delta\phi &= \delta \int k_i \, dx^i - h \, d\tau \\
&= \int \delta k_i \left(dx^i - \frac{\partial h}{\partial k_i} d\tau \right) - \int \left(dk_i + \frac{\partial h}{\partial x_i} d\tau \right) \delta x^i \\
&= 0
\end{aligned}
\qquad (2.105)
$$

where the integration is performed between two well-defined events in space–time. For arbitrary variations δk_i and δx^i, we can derive from here the photon canonical equations in covariant form:

$$\frac{dx^i}{d\tau} = \frac{\partial h}{\partial k_i}, \quad \frac{dk_i}{d\tau} = -\frac{\partial h}{\partial x^i}. \qquad (2.106)$$

In contrast with the above equations of motion (2.97), these new equations are formally analogous to the three-dimensional canonical ray equations (2.12). The question here is how to define the appropriate function $h(x^i, k_i)$ such that these new canonical equations are really equivalent to equations (2.97). This is simple for the plasma case, but not obvious for an arbitrary dielectric medium.

We notice that, for a plasma, the photon effective mass is not a function of the frequency. In this case, we can use the analogue of the covariant Hamiltonian for a particle with a finite rest mass:

$$h(x^i, k_i) = \frac{k_j k^j}{2m_{\text{eff}}} + \frac{1}{2} m_{\text{eff}} c^2. \qquad (2.107)$$

This is equivalent to

$$h(x^i, k_i) = \frac{(k_j k^j)c^2}{2\omega_p(x^i)} + \frac{1}{2}\omega_p(x^i)$$

$$= \frac{1}{2\omega_p(x^i)}\left[k^2 c^2 + \omega_p^2(x^i) - \omega^2\right]. \qquad (2.108)$$

It can easily be seen that, for $i = 0$, the use of this Hamiltonian function in equations (2.106) leads to the appropriate definition of the relativistic γ factor. For $i = 1, 2, 3$, it leads to equations (2.12), just showing that equations (2.106) are indeed an equivalent form of the photon canonical equations.

In the general case of a dielectric medium with refractive index n (where the photon effective mass is, in general, a function of the photon frequency), the covariant Hamiltonian $h(x^i, k_i) = 0$ has to satisfy the following conditions:

$$\frac{\partial h}{\partial \omega} = -\gamma, \qquad \frac{\partial h}{\partial t} = -\gamma\omega\frac{\partial}{\partial t}\ln n \qquad (2.109)$$

and

$$\frac{\partial h}{\partial \vec{k}} = \gamma\vec{v}, \qquad \frac{\partial h}{\partial \vec{r}} = -\gamma\omega\frac{\partial}{\partial \vec{r}}\ln n. \qquad (2.110)$$

For a non-dispersive medium, where $v = c/n$, we see that these four conditions are satisfied by the covariant Hamiltonian

$$h(x^i, k_i) = \gamma\left(\frac{kc}{n} - \omega\right) = \frac{\gamma c}{n(x^i)}\left[k_0 + \left(\sum_{i=1}^{3} k^i k_i\right)^{1/2}\right]. \qquad (2.111)$$

The photon ray theory for non-stationary plasmas was first proposed by us [66]. This work has been considerably extended in the present chapter. The covariant formulation was considered in reference [65], where the Hamiltonian (2.107) was first stated.

Chapter 3

Photon dynamics

Two different kinds of moving perturbation can be imagined in a non-stationary medium. The first one is a shock front or moving discontinuity, similar to the moving boundary between two media considered in the previous chapter. The second is a wave-type perturbation of the refractive index.

Both kinds of perturbation can be excited in a plasma by an intense laser pulse. They can also be analytically described in a simple way. The moving boundary will be the ionization front created by the laser pulse, and the wave perturbation will be the wakefield left behind the pulse.

For the sake of simplicity we will focus here on the dynamical properties of the photon motion in a non-magnetized plasma, in the presence of ionization fronts and of wakefields. At the end of this chapter, we will discuss the ways in which the same kind of perturbation can also be excited in magnetized plasmas and in other optical media.

As a particular example of photon acceleration in a dielectric medium, we will consider the well-known induced phase modulation processes.

3.1 Ionization fronts

One simple way of producing a moving discontinuity of the refractive index is to create an ionization front, which is the boundary between the neutral state and the ionized state of a given background gas. In other words, the ionization front is the boundary between a neutral and a plasma medium. The motion of this boundary is independent of the motion of the atoms, ions or electrons of the medium, which means that we can eventually produce a relativistic moving boundary in a medium where the particles stay nearly at rest.

The ionization front is one of the main problems in the theory of photon acceleration in plasmas, as referred to in chapter 1. A relativistic front can be efficiently produced by photoionization of a gas by an intense laser pulse [4, 47]. Because the photoionization is a fast process, occurring within a timescale of a

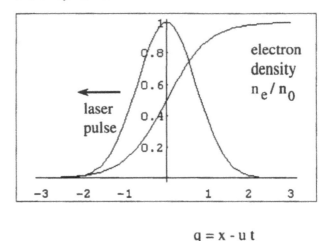

$$q = x - u\,t$$

Figure 3.1. Ionization front scheme: an intense laser pulse photoionizes a background gas. The resulting plasma boundary follows the pulse, moving at nearly the same speed.

few femtoseconds, the front will tend to replicate the form of the ionizing laser pulse and will move with nearly the same velocity.

The formation and propagation of such a front is a very complicated process. First of all, the laser pulse is moving at the boundary between two different media and, for that reason, its group velocity cannot be determined in a simple way.

Furthermore, its spectral content is also changing, due to the self-frequency shift, or acceleration of the laser photons by the laser itself, which will be discussed later. On the other hand, the photoionization process cannot be isolated from other ionization mechanisms because the primary electrons (created and subsequently accelerated by the laser field) will also contribute to the ionization process, by electron-impact ionization [60].

Here we are not interested in the details of the ionization front structure. It is only interesting to note that, behind the ionizing laser pulse, the plasma state will have a much longer lifetime than the duration of the pulse, because of the long timescales of the diffusion and recombination processes, which will eventually destroy it. We will therefore restrict our attention to the dynamics of individual probe or test photons, interacting with a given ionization front.

We start with a discussion of the different types of photon trajectory, as given by the one-dimensional ray equations

$$\frac{dq}{dt} = \frac{\partial \Omega}{\partial p}, \quad \frac{dp}{dt} = -\frac{\partial \Omega}{\partial q} \tag{3.1}$$

where $q = x - ut$, $p = k$ and u is the front velocity. We restrict our analysis to high-frequency photons, which is pertinent for laser propagation.

We can then neglect the contribution of the ion response and, for an isotropic or non-magnetized plasma, the Hamiltonian function Ω is simply determined by

$$\Omega \equiv \Omega(q, p, t) = \sqrt{p^2 c^2 + \omega_{p0}^2 f(q)} - up. \tag{3.2}$$

Here, ω_{p0} is the maximum value of the electron plasma frequency attained by the ionization front and $f(q)$ is a function increasing with q from 0 to 1 and describing the form of the front. One possible choice, already mentioned in chapter 2, is $f(q) = [1 + \tanh(k_f q)]/2$, where k_f determines the front width.

Another possible choice, for fronts created by a single step ionization of the background neutral atoms, by a Gaussian pulse, is $f(q) = \exp(-k_f^2 q^2)$, for $q < 0$, and $f(q) = 1$, for $q \geq 0$. Other, more complicated but more realistic forms of front profiles can also be proposed, but will not significantly modify our view of the various possible classes of photon trajectories.

From equation (3.2) we can determine the photon trajectory $p(q)$ in the two-dimensional phase space (q, p)

$$p(q) = \gamma^2 \frac{\Omega}{c} \left[s\beta \pm \sqrt{1 - \frac{1}{\gamma^2} \frac{\omega_{p0}^2}{\Omega^2} f(q)} \right] \tag{3.3}$$

where $\beta = |u|/c$, s is the sign of the velocity u and $\gamma^{-2} = (1 - \beta^2)$.

From this result we can establish two different conditions for photon reflection by the front. First, we can say that a reversal in direction of the photon trajectory in phase space (q, p) will occur if

$$\frac{\partial p}{\partial q} \to -\infty. \tag{3.4}$$

Using equation (3.3), we see that such a condition is satisfied if, for a given point $q = q_r$, we have

$$\Omega = \frac{\omega_{p0}}{\gamma} \sqrt{f(q_r)}. \tag{3.5}$$

Noting that the maximum value for $f(q)$ is equal to 1, we can say that the photon trajectory will be sooner or later reflected in phase space if $\Omega < \omega_{p0}/\gamma$. For a photon with initial frequency ω_1, which moves with a positive velocity along the x-axis (or, equivalently, along the q-axis) and then interacts with a front moving in the oposite direction (such that $s = -1$), we can write $\Omega = \omega_1(1 - s\beta) = \omega(1 + \beta)$.

This allows us to establish an upper limit ω_q for the frequencies of those photons with trajectories which reverse sense in phase space (q, p)

$$\omega_1 < \frac{\omega_{p0}}{1 + \beta} \sqrt{1 - \beta^2} \equiv \omega_q. \tag{3.6}$$

On the other hand, we have reflection in real space (a reflection observed in the laboratory frame of reference) if there is a turning point x where the photon momentum reverse changes its sign: $p = k = 0$. According to equation (3.3) this implies that

$$\Omega = \omega_{p0}\sqrt{f(q)}. \tag{3.7}$$

More generally, we can say that reflection in real space occurs if $\Omega < \omega_{p0}$. This allows us to establish a new upper limit ω_x for the frequency of photons with this kind of behaviour

$$\omega_1 < \frac{\omega_{p0}}{1 + \beta} \equiv \omega_x. \tag{3.8}$$

From equations (3.6, 3.8) we can get a simple relation between the two upper limits

$$\omega_x = \frac{\omega_q}{\sqrt{1 - \beta^2}}. \tag{3.9}$$

This shows that we will always have $\omega_x > \omega_q$. We are now ready to establish the distinct regimes of photon propagation across the ionization front or, in other words, the different types of trajectory in phase space. If we assume that the parameter ω_{p0} is fixed and if we vary the value of the incident photon frequency ω_1, we successively obtain:

(a) *type I trajectories*, for low-frequency photons, such that $\omega_1 < \omega_q$—there is reflection in both the real space and the phase space;
(b) *type II trajectories*, for moderate values of the frequency, such that $\omega_q < \omega < \omega_x$—there is reflection in real space but transmission (no reversal) in phase space. This means that the reflected photon and the front move both in the backward direction, but the front moves faster, which means that the photon can still cross the entire front region;
(c) *type III trajectories*, for high-frequency photons with frequencies such that $\omega_1 > \omega_x$—there is transmission in both the real space and the phase space.

If instead, we had assumed a fixed incident frequency ω_1 and had increased the value of the maximum plasma frequency of the front, we would have type III trajectories for $0 < \omega_{p0} < \omega_a$, where $\omega_a = \omega_1(1 + \beta)$, type II trajectories for moderate densities $\omega_a < \omega_{p0} < \omega_b$, where $\omega_b = \gamma\omega_a$, and type I trajectories for high-density trajectories, such that $\omega_{p0} > \omega_b$. These three types of trajectory are illustrated in figure 3.2.

The maximum values for the photon frequency shifts can easily be obtained from the invariance of Ω, as already shown in chapter 2, independently of the choice of the form function $f(q)$. For type I trajectories, we have

$$\omega_2 = \omega_1 \frac{1 + \beta}{1 - \beta} \tag{3.10}$$

where ω_2 is the final value of the frequency. This result is identical to that of reflection of an incident photon in a relativistic mirror with normalized velocity β.

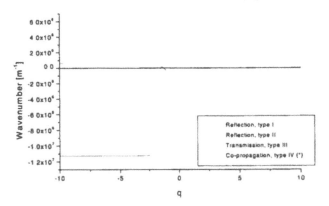

Figure 3.2. Trajectories of photons interacting with an ionization front. The three counter-propagating types of trajectory are shown for the following parameters: front velocity $\beta = 0.9$, electron density $n_e = 5 \times 10^{20}$ cm^{-3}. A co-propagating trajectory is also shown, but for a different electron density $n_e = 10^{20}$ cm^{-3}, for numerical reasons. This trajectory would not intersect the others if the plasma parameters were the same.

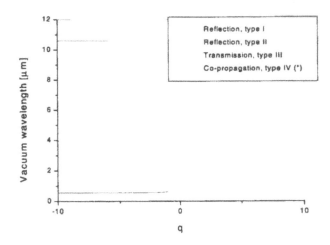

Figure 3.3. The same four trajectories as in figure 3.2, but represented in terms of the vaccum wavelength $\lambda = 2\pi c/\omega$, which is the prefered unit for the experimentalists to denote the measured photon frequency ω.

This is a very interesting result because, as we explained above, the particles of the media can stay at rest, and we still have a moving mirror effect.

The boundary between two different states of the medium (the neutral and the plasma state) act then as a moving material obstacle. Such an equivalence between the ionization front and a material moving mirror is however not complete, as discussed in chapter 4.

Using the same kind of argument, we can also easily obtain for type II and III trajectories the following frequency transformation:

$$\omega_2 = \frac{\omega_1}{1-\beta}\left[1 - \beta\sqrt{1 - (\omega_{p0}/\omega_1)^2(1-\beta)}\right]. \qquad (3.11)$$

For high-frequency photons, such that $\omega_1 \gg \omega_{p0}$, this leads to the following frequency shift:

$$\Delta\omega = \omega_2 - \omega_1 \simeq \frac{1}{2}\frac{\omega_{p0}^2}{\omega_1}\frac{\beta}{1+\beta}. \qquad (3.12)$$

As expected, the frequency shift for these highly energetic photons interacting with low-density fronts is very small. The interesting thing however is that even in this case a positive shift is expected. Actually, for $\beta \simeq 1$, the frequency shift is half of the value expected for the case of flash ionization ($\beta \to \infty$). This limiting case, and its physical meaning, will be discussed later.

Until now we have only considered photons counter-propagating with respect to the moving front. However, in order to get the full picture of the photon phase space, we have to consider *type IV trajectories*, corresponding to co-propagation. In this case, the frequency shift is

$$\omega_2 = \frac{\omega_1}{1-\beta}\left[1 - \beta\sqrt{1 - (\omega_{p0}/\omega_1)^2}\right]. \qquad (3.13)$$

For high frequencies $\omega_1 \gg \omega_{p0}$, this reduces to an expression similar to equation (3.12), but where $(1 + \beta)$ is replaced by $(1 - \beta)$, in the denominator. This means that, in co-propagation, much larger frequency shifts are expected.

Furthermore, two independent measurements of the frequency shifts, for co- and counter-propagation, will allow us to determine the two independent parameters of the ionization front: the normalized velocity β and the maximum plasma density or electron plasma frequency ω_{p0}.

Such a approach was followed in the experiments by Dias *et al* [21], who established a proof of principle of the photon accelerated process and also made the first observation of frequency shift in counter-propagation. Previous experiments of frequency up-shift by an ionization front propagating inside a microwave cavity, where essentially the co-propagating signal was observed, were reported by Savage *et al* [95]. Several other observations of frequency up-shift in microwave experiments were reported [52, 121].

These four types of trajectory correspond to four distinct regions in phase space, which are delimited by three separatrix curves. If we use $\Omega = \omega_{p0}/\gamma$ in equation (3.3), we obtain an expression for two of the separatrix curves

$$p(q) = \gamma\frac{\omega_{p0}}{c}\left[-\beta \pm \sqrt{1 - f(q)}\right]. \qquad (3.14)$$

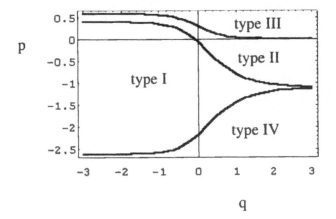

Figure 3.4. Photon interaction with ionization fronts: the three separatrix curves and the four types of trajectory.

If, instead, we use $\Omega = \omega_{p0}$, we get the third separatrix

$$p(q) = \gamma^2 \frac{\omega_{p0}}{c}\left[-\beta + \sqrt{1 - f(q)/\gamma^2}\right]. \tag{3.15}$$

These three curves, and the corresponding four regions of phase space, are shown in figure 3.4.

In the above discussion we have used an arbitrary form function $f(q)$ but where $f(q) = 0$ for $q \rightarrow -\infty$. Another possible version of the front assumes that the ionizing laser pulse is propagating not in a neutral gas, but in a partially ionized gas, with a residual plasma frequency equal to $\delta\omega_{p0}$, with $\delta \ll 1$. This model of the front was considered by Kaw *et al* [46]. It can be described by the following form function:

$$f(q) = \delta + \frac{1}{2}(1 - \delta)[1 + \tanh(k_f q)]. \tag{3.16}$$

This improvement in the model slightly changes the photon trajectories and the values of the resulting frequency shifts, but the basic picture of the photon phase space remains valid. For type I trajectories, the invariance of Ω now leads to the expression

$$\omega_2 = \frac{\Omega(\omega_1)}{1 - \beta^2}\left\{1 + \beta\sqrt{1 - \delta(1 - \beta^2)[\omega_{p0}/\Omega(\omega_1)]^2}\right\}. \tag{3.17}$$

Here, we have used $\Omega(\omega_1) = \omega_1[1 + \beta\sqrt{1 - \delta(\omega_{p0}/\omega_1)^2}]$. For $\delta = 0$ this reduces to equation (3.10), and, for $\delta \neq 0$ but for $\omega_1 \simeq \omega_{p0}$, this simplifies to

$$\omega_2 = \frac{\omega_1}{1 - \beta^2}\left[1 + \beta^2 + 2\beta\sqrt{1 - \delta}\right]. \tag{3.18}$$

Similarly, for type II and III trajectories, equation (3.11) is generalized to

$$\omega_2 = \frac{\Omega(\omega_1)}{1 - \beta^2} \left\{ 1 - \beta\sqrt{1 - (1 - \beta^2)\omega_{p0}/\Omega(\omega_1)^2} \right\} \qquad (3.19)$$

where $\Omega(\omega_1)$ is the same as in equation (3.17).

Finally, for the co-propagating type IV trajectories, we have, instead of equation (3.13),

$$\omega_2 = \frac{\Omega'(\omega_1)}{1 - \beta^2} \left\{ 1 + \beta\sqrt{1 - \delta(1 - \beta^2)[\omega_{p0}/\Omega'(\omega_1)]^2} \right\}. \qquad (3.20)$$

Here we have to use $\Omega'(\omega_1) = \omega_1[1 - \beta\sqrt{1 - (\omega_{p0}/\omega_1)^2}]$. Let us now turn to another (and eventually more realistic) model of the ionization front where, in the absence of the front, we have a neutral gas, but where the influence of the neutral gas on the refractive index n is not neglected as it was before.

If we consider a value $n \neq 1$ (but still neglect dispersion in the neutral region), we can use the following model:

$$\Omega(q, p) = \left\{ \frac{p^2 c^2}{1 + \chi[1 - f(q)]} + \omega_{p0}^2 f(q) \right\}^{1/2} - up. \qquad (3.21)$$

Here $\chi = n^2 - 1$ is the susceptibility of the background neutral gas, and a singly ionized plasma is assumed. With this model we still get the same picture of the photon phase space and similar expressions for the frequency shift. For instance, for type I trajectories, we have now

$$\omega_2 = \omega_1 \frac{1 + \beta n}{1 - \beta n}. \qquad (3.22)$$

For types II and III, we get

$$\omega_2 = \omega_1 \frac{1 + \beta n}{1 - \beta^2} \left\{ 1 - \beta \left[1 - \frac{\omega_{p0}^2}{\omega_1^2} \frac{(1 - \beta)^2}{(1 + \beta n)^2} \right]^{1/2} \right\} \qquad (3.23)$$

and, for type IV trajectories, we obtain

$$\omega_2 = \frac{\omega_1}{1 - \beta n} \left[1 - \beta\sqrt{1 - (\omega_{p0}/\omega_1)^2} \right]. \qquad (3.24)$$

Let us briefly consider oblique interaction of photons with an infinite ionization front. In this case, the generalization of the above one-dimensional analysis to the two-dimensional case allows us to determine the cut-off frequencies.

In particular, the condition for the existence of a type I trajectory (corresponding to reflection both in real space and in phase space) corresponds to incident photon frequencies ω_1 smaller than some limit ω_q, defined by

$$\omega_q = \frac{\omega_{p0}\sqrt{1-\beta^2}}{\sqrt{[1+\beta\cos(\theta_1)]^2 - (1-\beta^2)\sin^2(\theta_1)}} \tag{3.25}$$

where $\beta = |u|/c$.

For normal incidence, such that $\theta_1 = 0$, this expression reduces to equation (3.6). But a new and very interesting physical situation occurs when the incident photon is co-propagating in the vacuum region (or, more precisely, in the neutral gas region) and is overtaken by the ionization front.

Even if the photon travels with a velocity nearly equal to c, its velocity perpendicular to the front can be much smaller due to oblique propagation. In this case, we have $\pi/2 \leq \theta_1 \leq \pi$ and photon reflection can still take place, as shown by the above expression. This new effect can be called *co-propagating relativistic mirror* and it is due to oblique interaction with the front. The corresponding frequency shift can be obtained from the expressions of the two invariants I_1 and I_2, defined in chapter 2, and the result is

$$\omega_2 = \omega_1 \left[\frac{1 + \beta\cos(\theta_1) + \beta|\beta + \cos(\theta_1)|}{1 - \beta^2} \right]. \tag{3.26}$$

This effect was observed in the experiments by Dias *et al* [21] mentioned already.

3.2   Accelerated fronts

A front moving with constant velocity can only be considered as an ideal and limiting case. It is therefore quite natural to extend the above analysis to the more general situation of accelerated ionization fronts. This can easily be done by considering a plasma frequency space–time dependence of the form

$$\omega_p^2(\vec{r}, t) = \omega_p^2(\vec{r} - \vec{R}(t)). \tag{3.27}$$

For a density perturbation moving with constant velocity, we would simply have $\vec{R}(t) = \vec{v}t$, which is the case discussed in the previous section. But now we can have $d^2\vec{R}/dt^2 \neq 0$, which corresponds to accelerated density perturbations.

In order to treat our problem we can easily generalize the above canonical transformation $(\vec{r}, \vec{k}) \rightarrow (\vec{q}, \vec{p})$, using the following generating function:

$$F(\vec{r}, \vec{k}, t) = \vec{k} \cdot (\vec{r} - \vec{R}(t)). \tag{3.28}$$

This leads to

$$\vec{q} = \frac{\partial F}{\partial \vec{k}} = \vec{r} - \vec{R}(t), \quad \vec{p} = \frac{\partial F}{\partial \vec{r}} = \vec{k}. \tag{3.29}$$

The transformed Hamiltonian is

$$\Omega = \omega + \frac{\partial F}{\partial t} = \sqrt{p^2 c^2 + \omega_p^2(\vec{q})} - \vec{p} \cdot \frac{\partial \vec{R}}{\partial t}. \tag{3.30}$$

Now, the velocity of the front $\partial \vec{R}/\partial t$ is not constant, which means that Ω is no longer a constant of motion. New qualitative aspects of photon acceleration can be associated with accelerated fronts.

For co-propagation (photons and front propagating in the same direction) we can improve the frequency up-shifting by increasing the interaction time. For counter-propagation (photons and front propagating in the opposite sense) two sucessive reflections can take place.

We will illustrate these new aspects, by focusing our discussion on the one-dimensional problem. As before, we can write $\omega_p^2(q) = \omega_{p0}^2 f(q)$. But, instead of using our tanh model, we will assume an even simpler form for the ionization front, described by

$$\omega_p^2(q) = \begin{cases} 0 & (q < -1/k_f) \\ \omega_{p0}^2(1 + k_f q) & (-1/k_f < q < 0) \\ \omega_{p0}^2 & (q > 0) \end{cases}. \tag{3.31}$$

The width of the front is obviously equal to $1/k_f$. It is also clear that the photon frequency will only change as long as the photon travels inside the gradient region, or when the photon coordinates are such that

$$-1/k_f \quad < \quad q(t) \quad < \quad 0. \tag{3.32}$$

Inside that region, the photon frequency is determined by the expression

$$\omega(t) = \sqrt{p^2(t)c^2 - \omega_{p0}^2 k_f q(t)}. \tag{3.33}$$

In order to obtain an explicit expression for the frequency shift, we have to consider the photon equations of motion, allowing us to determine $q(t)$ and $p(t)$. They can be written as

$$\frac{dq}{dt} = \frac{pc^2}{\omega(t)} - \frac{\partial R}{\partial t} \tag{3.34}$$

$$\frac{dp}{dt} = \frac{1}{2} \frac{\omega_{p0}^2 k_f}{\omega(t)}. \tag{3.35}$$

These equations can easily be integrated by noticing that equation (3.33) allows us to write

$$q(t) = p^2(t) \frac{c^2}{\omega_{p0}^2 k_f} + Q(t) \tag{3.36}$$

with

$$Q(t) = -\frac{\omega^2(t)}{\omega_{p0}^2 k_f}.$$ (3.37)

Using equations (3.34, 3.35) we realize that the total time derivative of equation (3.36) leads to

$$\frac{dQ}{dt} = -\frac{\partial R}{\partial t}.$$ (3.38)

This can be integrated to give

$$Q(t) = Q_0 - R(t).$$ (3.39)

Going back to equations (3.33, 3.37), we obtain

$$\omega(t) = \sqrt{\omega_0^2 + \omega_{p0}^2 k_f R(t)}$$ (3.40)

where $\omega_0 \equiv \omega(t = 0)$ is the initial value of the photon frequency (just before entering the acceleration region).

The constant of integration in equation (3.39) is $Q_0 = -\omega_0^2/(\omega_{p0}^2 k_f)$, as can be seen from equation (3.37).

In the particular case of a front moving with constant velocity, we have $R(t) = v_f t$. This expression shows that the photon frequency grows with $\sqrt{t}$, as first noticed by Esarey *et al* [27]. This expression is valid as long as the photon stays inside the gradient region (3.32). It means that the closer the photon velocity is to the front velocity v_f, the longer t will be and the larger the final value of the frequency shift will be, as already shown in the previous section.

However, because of the photon acceleration process itself, the photon will have a tendency to escape from the gradient region, and a phase slippage between the photons and the front will always exist. Such a slippage can however be avoided (or at least significantly reduced) if the ionization front is accelerated as well.

Then, in principle, we can imagine an ideal situation where the photon co-propagating with the front is indefinitely accelerated up to arbitrary high frequencies, just by using a very small (but accelerated) plasma density perturbation. The external source responsible for the creation of such an accelerated plasma density perturbation would provide the energy necessary to accelerate the co-propagating photons.

Let us illustrate this idea with the simple case of a front moving with constant acceleration a:

$$R(t) = v_f t + \frac{1}{2}at^2.$$ (3.41)

In this case, we can estimate the maximum frequency up-shift for very underdense fronts (or, equivalently, for very high-frequency photons), such that $\omega_0^2 \gg \omega_{p0}^2$. The photon velocity can be assumed to be nearly equal to c and the total interaction time can be determined by the equality $ct_{in} - R(t_{in}) \simeq 1/k_f$. The solution is

$$t_{in} = \frac{c}{a}\left[(1 - \beta_f) - \sqrt{(1 - \beta_f)^2 + 2a/(k_f c^2)}\right] \qquad (3.42)$$

where we have used $\beta_f = |v_f|/c$. For small accelerations, such that $at_{in} \ll 2c(1 - \beta_f)$, we can simplify this expression and, using equation (3.40), we can write

$$\Delta\omega \equiv \omega(t_{in}) - \omega_0 \simeq \frac{\omega_{p0}^2}{2\omega_0^2}\frac{\beta_f}{1 - \beta_f}\left[1 + \frac{a}{2(1 - \beta_f)^2 c^2}\right]. \qquad (3.43)$$

The first term in this expression coincides with that derived in the previous section for fronts moving with constant velocity. The second one gives the contribution of the front acceleration. We can see that a positively accelerated front ($a > 0$) introduces a significant increase in the total frequency up-shift, especially due to the factor $(1 - \beta_f)^3$. This illustrates our previous statement that, by increasing the time of interaction between the photons and the front, we can optimize the photon acceleration process.

If we are not entirely satisfied with these qualitative arguments and want to replace them by a more detailed description of the photon dynamics, we can go back to equations (3.34, 3.35, 3.40) and derive explicit expressions for the photon trajectories. In particular, from equation (3.35), we can get

$$\frac{dp}{dt} = \frac{1}{2}\frac{\omega_{p0}\sqrt{k_f}}{\sqrt{-Q(t)}}. \qquad (3.44)$$

The photon trajectory will then be determined by

$$p(t) = p_0 + \frac{\omega_{p0}}{2}\sqrt{k_f}\int_0^t \frac{dt'}{\sqrt{R(t') - Q_0}}. \qquad (3.45)$$

For positively accelerated fronts ($a > 0$), this leads to

$$p(t) = p_0 + \omega_{p0}\sqrt{k_f/2a}\ln\left(\frac{\omega(t)\omega_{p0}\sqrt{k_f/2a}}{\omega_0 + \omega_{p0}v_k\sqrt{k_f/2a}}\frac{\partial R}{\partial t}\right). \qquad (3.46)$$

With equation (3.36), this completes the explicit integration of the photon trajectories. Until now we have focused our qualitative discussions on photons co-propagating with the front. But it can easily be seen that the above equations stay valid for the counter-propagating photons, except that in equations (3.42, 3.43) the factor $(1 - \beta_f)$ is replaced by $(1 + \beta_f)$. The frequency shift is still increased by a positive front acceleration.

However, a qualitatively new effect occurs for co-propagating photons: more than one turning point, defined by

$$\frac{dq}{dt} = 0 \qquad (3.47)$$

can be observed along the same trajectory.

This means that a counter-propagating photon can be first reflected by the front, suffering the corresponding double Doppler shift, and then it is caught again by the front, and reflected a second time into the co-propagating direction. These doubly reflected trajectories were numerically identified by Silva [99] and can attain considerably large frequency shifts.

3.3   Photon trapping

3.3.1   Generation of laser wakefields

It is well known that a plasma can support several kinds of electrostatic wave and oscillations. One of the basic types of such waves is called an electron plasma wave, because its frequency is nearly equal to the electron plasma frequency [82, 108]. To be more precise, the electron plasma waves are characterized by the following dispersion relation:

$$\omega^2 = \omega_p^2 + 3k^2 v_{the}^2 \tag{3.48}$$

where ω and k are here the frequency and wavenumber of the electrostatic oscillations, and $v_{the} = \sqrt{T_e/m}$ is the electron thermal velocity and T_e the electron temperature.

The electron plasma waves can only persist in a plasma if their wavenumber satisfies the inequality $k \ll \omega_p/v_{the}$. Otherwise, they will be strongly attenuated by electron Landau damping.

Such a damping is due to the energy exchange between the waves and the electrons travelling with a velocity nearly equal to the wave phase velocity $v_\phi = \omega/k$ (the so-called resonant electrons). This is equivalent to saying that the electron plasma waves can only exist if their wavelength is much larger than a characteristic scale length of the plasma $\lambda \ll \lambda_{De}$, where $\lambda_{De} = v_{the}/\omega_p$ is called the electron Debye length.

The existence of the electron Landau damping implies that the frequency of the allowed electron plasma waves is always very close to the electron plasma frequency, $\omega \sim \omega_p$. As can be seen from the above dispersion relation, the phase velocity of the electron plasma waves can be arbitrarily high: $v_\phi v_g = 3v_{the}^2$.

It is then possible to excite electrostatic waves with relativistic phase velocities $v_\phi \simeq c$, and nearly zero group velocities. These waves will not be Landau damped, because resonant electrons will be nearly absent (except for extremely hot plasmas).

Relativistic electron plasma waves can be excited by intense laser beams in a plasma. Using a simplified but nevertheless accurate view of this problem, we can say that there are two different excitation mechanisms: the beat-wave and the wakefield.

The first mechanism [64, 77, 92] is valid for long laser pulses, with a duration Δt much larger than the period of the electron plasma wave: $\Delta t \gg \omega_p^{-1}$. In this

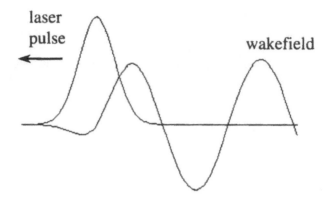

Figure 3.5. Schematic of the laser wakefield mechanism. A short laser pulse propagating in a plasma excites a relativistic electron plasma wave.

case, the superposition of two parallel laser pulses with different but very close frequencies, ω_1 and ω_2, such that their difference matches the electron plasma frequency, $\omega_1 - \omega_2 \simeq \omega_p$ can resonantly excite an electron plasma wave. The phase velocity of this wave is determined by the laser beating $v_\phi = (\omega_1 - \omega_2)/(k_1 - k_2)$.

The second mechanism [12, 34, 89] is valid for short laser pulses, with a duration of the order of the electron plasma period $\Delta t \simeq \omega_p^{-1}$. A single laser pulse can then excite a tail of electron plasma oscillations, which is usually called the laser wakefield. It is important to notice that the phase velocity of this wakefield is nearly equal to the group velocity of the laser pulse.

With both the beat-wave and the wakefield mechanisms, relativistic electron plasma waves can be excited. Interaction of these waves with fast electrons can accelerate them further and provide the basis for a new generation of particle accelerators [109].

The problem of electron acceleration is not our concern here. Our interest is mainly focused on the analogies between charged particles and photons and, especially, on the possibility of accelerating and trapping photons in the field of an electron plasma wave. But, it should also be mentioned that the study of photon acceleration and trapping by a relativistic plasma wave can be an important diagnostic tool for particle accelerator research [22, 107].

3.3.2   Nonlinear photon resonance

From now on, we will simply refer to a relativistic electron plasma wave as a wakefield. In order to describe such a wakefield, we will assume a plasma frequency perturbation of the form

$$\omega_p^2(\vec{q}) = \omega_{p0}^2 \left[1 + \epsilon \cos(\vec{k}_p \cdot \vec{q}) \right] \tag{3.49}$$

where $\epsilon < 1$ is the relative amplitude of the wakefield and $\vec{k}_p$ is the wakefield wavevector, and the subscript p is added in order to avoid confusion with the photon wavevector.

The photon equations of motion can be explicitly written as

$$\frac{d\vec{q}}{dt} = \frac{c^2 \vec{p}}{\omega(\vec{q}, \vec{p})} - \vec{u} \tag{3.50}$$

$$\frac{d\vec{p}}{dt} = \frac{\epsilon}{2} \frac{\omega_{p0}^2 \vec{k}_p}{\omega(\vec{q}, \vec{p})} \sin(\vec{k}_p \cdot \vec{q}) \tag{3.51}$$

where $\vec{u} \simeq (\omega_{p0}/|\vec{k}_p|^2)\vec{k}_p$ is the wakefield phase velocity and

$$\omega(\vec{q}, \vec{p}) = \left\{ p^2 c^2 + \omega_{p0}^2[1 + \epsilon \cos(\vec{k}_p \cdot \vec{q})] \right\}^{1/2}. \tag{3.52}$$

Let us determine the fixed points, defined by

$$\frac{d\vec{q}}{dt} = 0, \quad \frac{d\vec{p}}{dt} = 0. \tag{3.53}$$

This is equivalent to writing

$$c^2 \vec{p} = \omega(\vec{q}, \vec{p})\vec{u}, \quad \sin(\vec{k}_p \cdot \vec{q}) = 0. \tag{3.54}$$

At this point we could split the variables $\vec{q}$ and $\vec{p}$ into their parallel and perpendicular components with respect to the direction of the wakefield propagation $\vec{u}/u$, or $\vec{k}_p/k_p$. However, it can easily be realized that the perpendicular components play only a secondary role in the photon dynamics.

We will now proceed by just retaining the simple case of one-dimensional motion, where $q_\parallel = q$, $p_\parallel = p$, and $q_\perp = p_\perp = 0$. From equations (3.53) we can then explicitly establish the elliptic fixed points as

$$q_e = \frac{\pi}{k_p}, \quad p_e = s\gamma\beta\frac{\omega_{p0}}{c}\sqrt{1 - \epsilon} \tag{3.55}$$

and the hyperbolic fixed points as

$$q_h = 0, \frac{2\pi}{k_p}, \quad p_h = s\gamma\beta\frac{\omega_{p0}}{c}\sqrt{1 + \epsilon}. \tag{3.56}$$

Here we have used $\beta = |u|/c$, $s =$ sign of u and $\gamma = (1 - \beta^2)^{-1/2}$. We should notice that, in contrast with the simple pendulum, the momenta p_e and p_h

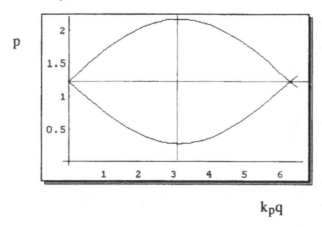

$$k_pq$$

Figure 3.6. The separatrix curves of the photon nonlinear resonance in phase space (q, p).

corresponding to the elliptic and hyperbolic fixed points are not identical. The photon nonlinear resonance described by equations (3.50, 3.51) appears then as a kind of asymmetric pendulum [98].

Furthermore, the resonance asymmetry increases with the wakefield relativistic γ-factor. For small wakefield amplitudes $\epsilon \ll 1$, the distance between the two fixed points can be written as

$$|p_e - p_h| \simeq \gamma \beta \frac{\omega_{p0}}{c} \epsilon. \tag{3.57}$$

A qualitative representation of the photon nonlinear resonance in phase space (q, p) is shown in figure 3.6.

It is also interesting to notice that, for the one-dimensional case, the invariant Ω can be written as

$$\Omega = \omega(q, p) - up = \gamma^{-2}\omega(q, p). \tag{3.58}$$

This means that the elliptic fixed point given by equation (3.55) can be defined by

$$p_e = s\gamma^2 \beta \frac{\Omega_e}{c^2} \tag{3.59}$$

where

$$\Omega_e = \frac{\omega_{p0}}{\gamma}\sqrt{1 - \epsilon}. \tag{3.60}$$

The value of p_h, for the hyperbolic fixed point, is also determined by an expression similar to equation (3.59), where Ω_e is replaced by Ω_h such that

$$\Omega_h = \frac{\omega_{p0}}{\gamma}\sqrt{1 + \epsilon}. \tag{3.61}$$

Replacing this expression in the definition of the invariant Hamiltonian Ω, we obtain the equation for the separatrix

$$(\Omega_h - up)^2 = p^2 c^2 + \omega_p^2(q), \tag{3.62}$$

or, more explicitly,

$$p = \gamma \frac{\omega_{p0}}{c} \sqrt{1+\epsilon} \left[s\beta \pm \sqrt{1 - \frac{1 + \epsilon \cos(k_p q)}{1 + \epsilon}} \right]. \tag{3.63}$$

For $q = 0, (2\pi/k_p)$, this reduces to $p = p_h$. On the other hand, for $q = (\pi/k_p)$, this expression determines the maximum and the minimum values of the photon momentum on the separatrix

$$p_\pm = \gamma \frac{\omega_{p0}}{c} \sqrt{1+\epsilon} \left[s\beta \pm \sqrt{\frac{2\epsilon}{1+\epsilon}} \right]. \tag{3.64}$$

The width of the nonlinear resonance will then be given by the difference between these two extreme values of the separatrix curves:

$$\Delta p_{max} = p_+ - p_- = 2\gamma \frac{\omega_{p0}}{c} \sqrt{2\epsilon}. \tag{3.65}$$

From this analysis we recognize that the photons are trapped by the electron plasma wave if their trajectories are associated with values of the invariant Hamiltonian Ω such that

$$\Omega_e \leq \Omega < \Omega_h. \tag{3.66}$$

Inside the separatrix, the photon frequency will oscillate according to the value of Ω characterizing its trajectory. We will have no frequency variation only for exactly resonant photon trajectories such that $p = p_e$, or $\Omega = \Omega_e$, corresponding to the elliptic fixed point.

The frequency variation will grow when we approach the separatrix curve, but such a change will require longer and longer times to take place. We will also attain a maximum for the frequency variation at the separatrix, between the points p_+ and p_- defined by equation (3.64), if we wait an infinite time.

For very high-frequency photons, such that $\omega \gg \omega_{p0}$, we have $\omega \simeq pc$ and the maximum frequency shift will be determined by $\Delta\omega_{max} \simeq c\Delta p_{max} = 2\gamma\omega_{p0}\sqrt{2\epsilon}$. For untrapped photons, the photon frequency will also oscillate, but with smaller and smaller amplitudes when we displace the photon trajectories away from the separatrix. This is illustrated in figure 3.7.

It is now interesting to look at the deeply trapped trajectories, oscillating around $p = p_e$. If we introduce $q = (\pi/k_p) + \tilde{q}$ and linearize the photon equations of motion around the elliptic fixed point, we can easily obtain

$$\frac{d^2\tilde{q}}{dt^2} + \omega_b^2 \tilde{q} = 0 \tag{3.67}$$

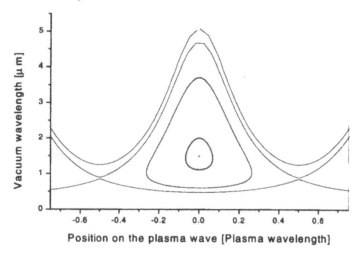

Figure 3.7. Examples of photon trajectories around a nonlinear resonance, showing trapped and untrapped motion, corresponding to the following parameters: $\beta = 0.9995$, $n_e = 10^{18} \text{cm}^{-3}$, $\epsilon = 0.5$. The photon frequency ω is represented in terms of the vacuum wavelength $\lambda = 2\pi c/\omega$.

which corresponds to a linear oscillator with a frequency

$$\omega_b = \frac{ck_p}{\gamma}\sqrt{\frac{\epsilon}{2(1-\epsilon)}}. \tag{3.68}$$

This can be called the *photon bounce frequency*. It is instructive to compare its value with the bounce frequency of an electron trapped in the field of the same electron plasma wave [82]

$$\omega_{be} = \sqrt{(e/m)E_0 k_p} \tag{3.69}$$

where E_0 is the amplitude of the electric field associated with this wave.

Using Poisson's equation we can relate it to the electron density amplitude perturbation

$$k_p E_0 = \frac{e\tilde{n}}{\epsilon_0} = \frac{en_0}{\epsilon_0}\epsilon. \tag{3.70}$$

This allows us to write the electron bounce frequency as

$$\omega_{be} = \omega_{p0}\sqrt{\epsilon}. \tag{3.71}$$

This shows the striking similarities of the photons and electrons trapped in the field of an electron plasma wave. Other and eventually even more surprising similarities will be discussed later.

3.3.3   Covariant formulation

As an example of an application of the covariant equations established in section 2.5, we can describe the photon trapping by using equations (2.106, 2.108) with

$$\omega_p^2(x^i) = \omega_{p0}^2\left[1 + \epsilon\cos(K_i x^i)\right] \tag{3.72}$$

where K^i is the four-momentum associated with the plasma wakefield and ϵ is, as before, its amplitude. Let us write

$$K_i x^i = K^1 x^1 - K^0 x^0 = k' x^1 - \omega' t \tag{3.73}$$

where $\omega' \simeq \omega_{p0}$ is the wakefield frequency and is related to k' by the dispersion relation (3.48).

In this particular case, the covariant Hamiltonian (2.108) is independent of x^2 and x^3, which means that the components k^2 and k^3 of the photon momentum are two constants of motion. We will assume a one-dimensional photon propagation by making these two constants equal to zero: $k^2 = k^3 = 0$. We can then write the reduced one-dimensional form of the covariant Hamiltonian as

$$h(x^0, x^1, k_0, k_1) = \frac{k_1^2 - k_0^2}{2\omega_p(x^0, x^1)}c^2 + \frac{1}{2}\omega_p(x^0, x^1). \tag{3.74}$$

The corresponding equations of motion in the relativistic space–time are

$$\frac{dx^0}{d\tau} = \frac{\partial h}{\partial k_0} = -\frac{k_0 c^2}{\omega_p(x^0, x^1)} \tag{3.75}$$

$$\frac{dx^1}{d\tau} = \frac{\partial h}{\partial k_1} = \frac{k_1 c^2}{\omega_p(x^0, x^1)} \tag{3.76}$$

and

$$\frac{dk_0}{d\tau} = -\frac{\partial h}{\partial x^0} = -\frac{1}{2}\left[1 - \frac{k_1^2 - k_0^2}{\omega_p^2(x^0, x^1)}c^2\right]\frac{\partial\omega_p}{\partial x^0} \tag{3.77}$$

$$\frac{dk_1}{d\tau} = -\frac{\partial h}{\partial x^1} = -\frac{1}{2}\left[1 - \frac{k_1^2 - k_0^2}{\omega_p^2(x^0, x^1)}c^2\right]\frac{\partial\omega_p}{\partial x^1} \tag{3.78}$$

where τ is the photon proper time.

In order to write these equations in a more convenient form, it is useful to introduce new adimensional variables, such that

$$x = K^1 x^1 = k' x^1, \quad y = k' x^0 \tag{3.79}$$

and

$$v = \frac{k_1 c}{\omega_{p0}}, \quad u = \frac{k_0 c}{\omega_{p0}}. \tag{3.80}$$

We can see from the definition of the variable y that $K^0 x^0 = \beta' y$, where $\beta' = \omega'/k'c$ is the normalized phase velocity of the wakefield. The normalized proper time z, and Hamiltonian h_0, are defined by

$$dz = k'c\,d\tau, \quad h_0 = \frac{h}{\omega_{p0}}. \tag{3.81}$$

We can state the new equations of motion as

$$\frac{dx}{dz} = \frac{\partial h_0}{\partial v}, \quad \frac{dy}{dz} = \frac{\partial h_0}{\partial u} \tag{3.82}$$

and

$$\frac{dv}{dz} = -\frac{\partial h_0}{\partial x}, \quad \frac{du}{dz} = -\frac{\partial h_0}{\partial y}. \tag{3.83}$$

The explicit form of the normalized Hamiltonian is

$$h_0(x, v; y, u) = \frac{v^2 - u^2 + 1 + \epsilon \cos(x - \beta' y)}{2\sqrt{1 + \epsilon \cos(x - \beta' y)}}. \tag{3.84}$$

By definition, we always have $h_0 = 0$. Noting that the denominator in this equation cannot become imaginary because the wakefield modulation is always less than one, $\epsilon < 1$, we can write

$$u^2 = v^2 + 1 + \epsilon \cos(x - \beta' y). \tag{3.85}$$

This is nothing but the photon dispersion equation, written in adimensional form. Noticing that h_0 only depends on x and y through $(x - \beta' y)$, we conclude that

$$\frac{\partial h_0}{\partial x} = -\frac{1}{\beta'}\frac{\partial h_0}{\partial y}. \tag{3.86}$$

Replacing this in the canonical equations (3.83), we obtain

$$\frac{dv}{dz} = -\frac{1}{\beta'}\frac{du}{dz}. \tag{3.87}$$

This expression means that, apart from the Hamiltonian H_0, it is possible to define another constant of motion I, such that

$$I = u + \beta' v = \text{const.} \tag{3.88}$$

The existence of these two independent constants of motion, h_0 and I, proves that the motion of a photon in a sinusoidal wakefield is integrable, as expected. Notice that this equation is equivalent to stating that $\omega - kv'$ is a constant, where $v' = \omega'/k'$ is the phase velocity of the wakefield. This invariant was already identified in the non-covariant formulation of photon acceleration. It will appear again in the full wave description, where it corresponds to the wave phase invariance.

The position of the nonlinear resonance contained in the Hamiltonian (3.84) is determined by the stationary condition (in the proper time z), or

$$\frac{d}{dz}(x - \beta'y) = 0. \tag{3.89}$$

Noting that, from the canonical equations (3.82), we have

$$\frac{dx}{dz} = \frac{v}{\Omega_p}, \quad \frac{dy}{dz} = -\frac{u}{\Omega_p} \tag{3.90}$$

where $\Omega_p = \sqrt{1 + \epsilon \cos(x - \beta'y)}$ is a positive quantity, we can write this stationary condition as

$$v + \beta'u = 0. \tag{3.91}$$

In non-normalized variables this is equivalent to $kc^2/\omega = \omega'/k'$. This means that the nonlinear resonance corresponds to the equality between the photon group velocity $v_g \equiv kc^2/\omega$ and the phase velocity of the wakefield perturbation ω'/k'.

Replacing this resonance condition in equation (3.85), we obtain

$$u^2(1 - \beta') = 1 + \epsilon \cos(x - \beta'y). \tag{3.92}$$

The centre of this resonance corresponds to $(x - \beta'y) = \pi/2$, which leads to the following coordinates:

$$u_0 = -\frac{1}{\sqrt{1 - \beta'^2}} \equiv -\gamma', \quad v_0 = \frac{\beta'}{\sqrt{1 - \beta'^2}} = -\beta'u_0. \tag{3.93}$$

The value of the invariant I, defined by equation (3.88), corresponding to this particular photon trajectory is equal to $I_0 = -\sqrt{1 - \beta'}$. We can also see, from equation (3.92), that the maximum excursion of u^2 associated with trapped trajectories inside the nonlinear resonance is determined by $\delta u^2(1 - \beta'^2) = 2\epsilon$.

The resonance half-width is then given by

$$\delta u_0 = \sqrt{\frac{\epsilon}{2}} \frac{1}{\sqrt{1 - \beta'}}. \tag{3.94}$$

This completes the characterization of the trapped photon trajectories in the four-dimensional relativistic space–time.

3.4   Stochastic photon acceleration

In his famous experiment, Isaac Newton inaugurated spectroscopy by decomposing white light into its spectral components with the help of a prism. In very recent days it was discovered that the opposite process can also occur and, starting from a nearly monochromatic spectral light source, we can regenerate white light. For that purpose, stochastic acceleration of photons can be used.

3.4.1   Motion in two wakefields

Let us first discuss the simple case of photon motion in the presence of two different wakefields (or, in other words, two relativistic electron plasma waves) with amplitudes ϵ_1 and ϵ_2, distinct wavevectors $\vec{k}_1$ and $\vec{k}_2$ and distinct phase velocities $\vec{u}_1 = (\omega'/|k_1|^2)\vec{k}_1$ and $\vec{u}_2 = (\omega'/|k_2|^2)\vec{k}_2$.

The photon dispersion relation can now be written as

$$\omega = \left\{ k^2 c^2 + \omega_{p0}^2 \left[1 + \sum_{i=1,2} \epsilon_i \cos(\vec{k}_i \cdot \vec{r} - \omega_i t) \right] \right\}^{1/2} . \tag{3.95}$$

Using the canonical variables $\vec{q} = \vec{r} - \vec{u}_1 t$ and $\vec{p} = \vec{k}$, we can write the photon equations of motion in the form of equations (3.1), with the new Hamiltonian (which is now time dependent)

$$\Omega = \omega - \vec{u}_1 \cdot \vec{p} = \left\{ p^2 c^2 + \omega_{p0}^2 [1 + \epsilon_1 \cos(\vec{k}_1 \cdot \vec{q}) \right.$$
$$\left. + \epsilon_2 \cos(\vec{k}_2 \cdot (\vec{q} - \vec{v}t))] \right\}^{1/2} - \vec{u}_1 \cdot \vec{p} \tag{3.96}$$

where

$$\vec{v} = \vec{u}_1 - \vec{u}_2 \simeq \omega_{p0} [(\vec{k}_1/k_1^2) - (\vec{k}_2/k_2^2)]. \tag{3.97}$$

In order to simplify the discussion, we will assume that the wakefields propagate in the same direction, and we will concentrate on the one-dimensional case. The extension to three dimensions is straightforward and does not lead to qualitatively new effects.

We can see from the above Hamiltonian (or from its one-dimensional version) that two nonlinear resonances exist in the (q, p) phase plane. They are defined by the elliptic fixed points

$$q_i = \frac{\pi}{k_p}, \qquad p_i = s_i \gamma_i \beta_i \frac{\omega_{p0}}{c} \sqrt{1 - \epsilon_i} \tag{3.98}$$

for $i = 1, 2$. The following definitions were used here: s_i = sign of u_i, $\beta_i = |u_i|/c$ and $\gamma_i = (1 - \beta_i^2)^{-1/2}$.

It is also quite well known that the second of these two resonances (the one corresponding to $i = 2$) can only be seen in the phase plane if we use a stroboscopic plot of the photon motion, at the instants $t_n = (2n\pi/k_2 v)$, for n integer. Such a discrete representation of the motion is usually called a Poincaré map.

In accordance with equation (3.65) we can also say that the the width of these two resonances is determined by

$$(\Delta p)_i = 2\gamma_i \frac{\omega_{p0}}{c} \sqrt{2\epsilon_i}. \tag{3.99}$$

It is well known from the theory of dynamical systems [59, 124] that the interaction between the two resonances leads to the destruction of the separatrix curves. These curves will both break up into a thin region of stochastic motion and the width of such a region will exponentially grow with the amplitude of the resonances ϵ_i. With increasing values of ϵ_i, the two stochastic regions will eventually merge, and a large fraction of the photon phase space (q, p) will be filled with stochastic trajectories, leading to what is called large-scale stochasticity.

Occurrence of large-scale stochastic acceleration is dictated by a qualitative criterion, called the Chirikov criterion or the resonance overlapping criterion. This is confirmed by numerical calculations within an error of a few per cent, for a large variety of similar nonlinear Hamiltonian motions. The criterion states that large-scale stochasticity occurs when the sum of the resonance half-widths becomes larger than the distance between these resonances.

In our case it can be written as

$$(\Delta p)_1 + (\Delta p)_2 \geq 2|p_2 - p_1|. \tag{3.100}$$

Using equations (3.98, 3.99) we can write the overlapping criterion as

$$\frac{\gamma_1\sqrt{2\epsilon_1} + \gamma_2\sqrt{2\epsilon_2}}{|s_2\beta_2\gamma_2\sqrt{1-\epsilon_2} - s_1\beta_1\gamma_1\sqrt{1-\epsilon_1}|} \geq 1. \tag{3.101}$$

It is important to notice that this criterion is independent of the mean electron plasma frequency ω_{p0}, and that it only depends on the velocities and amplitudes of the two wakefields. If we assume that the two amplitudes are equal, $\epsilon_1 = \epsilon_2 = \epsilon$, we can simplify the overlapping criterion and write

$$\sqrt{\frac{2\epsilon}{1-\epsilon}} \frac{1+v}{|s_2\beta_2 - s_1\beta_1 v|} \geq 1 \tag{3.102}$$

where $v = \gamma_1/\gamma_2$.

In the weakly relativistic limit, where we have $\gamma_i \simeq 1$ and $v \simeq 1$, we conclude from this expression that large-scale stochasticity occurs for

$$\epsilon \geq \frac{1}{2^3}(s_2\beta_2 - s_1\beta_1)^2. \tag{3.103}$$

Noting that β_i are always smaller than one, we conclude that this criterion is compatible with low values of the wakefield amplitude $\epsilon \ll 1$. In the opposite limit of strongly relativistic phase velocities of the plasma wakefields, where $\gamma_i \gg 1$ and $\beta_i \simeq 1$, we obtain from equation (3.102)

$$\epsilon \geq \frac{1}{2^3}(1 - v)^2 \tag{3.104}$$

where we have assumed that the wakefields propagate in the same direction, $s_1 = s_2$, and that their relativistic γ factors are similar, $v \simeq 1$.

We see that, again, the criterion is compatible with low wakefield amplitudes $\epsilon \ll 1$. Even in the most unfavourable situation of very different relativistic factors, $\nu \gg 1$, we get from equation (3.103) $\epsilon \geq 1/3$. We can then say that, in very broad terms, both in the weakly and the strongly relativistic limits, transition from regular to stochastic trajectories can be expected with moderately low wakefield amplitudes.

However, the strongly relativistic case is physically more interesting, because the width of the region between the two resonances where stochastic acceleration can occur is proportional to the relativistic gamma factors, and a larger frequency spread of a bunch of photons with nearly identical initial frequencies can be obtained. From initial monochromatic radiation we can then obtain broadband radiation (white light) [66].

A similar situation occurs if, instead of having two different wakefields, we can excite a single wakefield in a plasma, but with a modulated amplitude. Such a modulation can be due, for instance, to the existence of ion acoustic waves propagating in the background plasma. Here, instead of two nonlinear resonances, we have three nearby resonances to which we can apply the same overlapping criterion. This was studied, using the covariant formulation, in reference [65].

3.4.2 Photon discrete mapping

It is also interesting to consider the interaction of a photon moving in a plasma with an electrostatic wavepacket containing a large spectrum of electron plasma waves. Here we will follow a procedure similar to that used by Zaslavski *et al* [125] for studying charged particles.

The photon dispersion relation (3.95) is now replaced by

$$\omega = \left\{ k^2 c^2 + \omega_{p0}^2 \left[1 + \sum_{n=-\infty}^{\infty} \epsilon_n \cos(\vec{k}_n \cdot \vec{r} - \omega_n t) \right] \right\}^{1/2} . \tag{3.105}$$

The corresponding one-dimensional photon equations of motion are

$$\frac{dx}{dt} = \frac{kc^2}{\omega}, \quad \frac{dk}{dt} = \frac{\omega_{p0}^2}{2\omega} \sum_{n=-\infty}^{\infty} \epsilon_n k_n \sin(k_n x - \omega_n t). \tag{3.106}$$

As a simple and reasonable model for the wavepacket, we can use

$$\epsilon_n k_n = a, \quad \omega_n = \omega_0 \simeq \omega_{p0}, \quad k_n = k_0 + n\Delta k \tag{3.107}$$

where a is a constant and $\Delta k \ll k_0$ is the characteristic distance between two consecutive spectral components.

This model is convenient from the physical point of view because, in contrast with similar models used in nonlinear dynamics where the amplitudes are assumed constant, it corresponds to a spectrum with variable amplitudes, where the lower amplitudes correspond to the higher wavenumbers. This is compatible with the idea that electron Landau damping of the wakefield spectrum prevents the higher wavenumbers.

The sum in equation (3.106) can then be replaced by

$$a \sin \theta \sum_{n=-\infty}^{\infty} \cos(n\Delta kx) \tag{3.108}$$

where $\theta = k_0 x - \omega_0 t$ is the phase of the central spectral component. We can also use the identity

$$\sum_{n=-\infty}^{\infty} \cos(n\Delta kx) = \frac{2\pi}{\Delta k} \sum_{n=-\infty}^{\infty} \delta(x - x_n) \tag{3.109}$$

with $x_n = (2n\pi/\Delta k) \equiv nL$. This means that, from equation (3.106), we can write

$$\frac{dk}{dt} = \frac{\omega_{p0}^2}{2\omega} aL \sin \theta \sum_{n=-\infty}^{\infty} \delta(x - x_n). \tag{3.110}$$

This shows that the photon moves with a constant wavenumber (or a constant velocity), except when it crosses the point $x = x_n$, where it suffers a sudden kick and abruptly changes its wavenumber. This new wavenumber remains constant until it crosses the next point $x = x_{n+1}$, and so on.

This picture of the photon motion means that we can transform the variable position x inside the delta function argument into a time variable, just by using the relation $x = |v_g|t = (|k|c^2/\omega)t$. The above equation is then replaced by

$$\frac{dk}{dt} = \frac{\omega_{p0}^2}{2|k|c^2} aL \sin \theta \sum_{n=-\infty}^{\infty} \delta(t - t_n). \tag{3.111}$$

The instants t_n are defined by $x(t_n) = x_n$, apart from a minor detail resulting from the fact that the photon can eventually be reflected at some point x_n, and in that case the instant t_{n+1} will be defined by $x(t_{n+1}) = x_{n-1}$. Of course, such a reversal in the direction of photon propagation is not likely to occur for energetic photons, with initial frequencies much larger than the electron plasma frequency.

Let us now introduce an adimensional variable w, such that

$$w = \frac{kc^2}{\omega_{p0}^2} |k| = \frac{k^2 c^2}{\omega_{p0}^2} s_k \tag{3.112}$$

where s_k is the sign of k. This is equivalent to writing

$$k = \frac{\omega_{p0}}{c} |w|^{1/2} s_w \tag{3.113}$$

where s_w is the sign of w. According to equation (3.111) the new variable w will evolve in time as

$$\frac{dw}{dt} = aL\sin\theta \sum_{n=-\infty}^{\infty} \delta(t-t_n). \qquad (3.114)$$

This has to be completed with an equation describing the time evolution for the phase variable θ. By definition, we have $d\theta/dt = (k_0c^2/\omega)k - \omega_0$. But, at this point, we should notice that it is physically quite natural to assume that $\epsilon_n \ll 1$.

According to equation (3.105), this allows us to replace ω by $\bar{\omega}$ and to write, using the new variable w

$$\frac{d\theta}{dt} = k_0c\frac{\omega_{p0}}{\bar{\omega}}s_w - \omega_0. \qquad (3.115)$$

This equation, with $\bar{\omega}$ in the denominator, is also physically very convenient because it guarantees that θ varies continuously across the points $t = t_n$. We now have two coupled and closed equations for the variables w and θ, equations (3.114, 3.115), and we can build up a map on the new phase plane (w, θ). For that purpose, we define

$$w_n = w(t_n^-), \quad \theta_n = \theta(t_n^-) \qquad (3.116)$$

where $t_n^\pm = t_n \pm \delta$, with $\delta \to 0$, represent the instants imediately before $(-)$ and after $(+)$ the critical instants t_n.

Equations (3.114, 3.115) show that, at $t = t_n$, the variable w suffers a sudden jump, and that the phase θ remains constant. This can be stated as

$$w(t_n^+) - w(t_n^-) = aL\sin\theta, \quad \theta(t_n^+) - \theta(t_n^-) = 0. \qquad (3.117)$$

During the interval (t_n^+, t_{n+1}^-) the variable w, and the corresponding photon momentum p, remain unchanged. This is equivalent to stating that $w(t_{n+1}^-) = w(t_n^0)$, or

$$w_{n+1} = w_n + aL\sin\theta_n. \qquad (3.118)$$

We can also see from equation (3.115) that the time derivative of the phase θ also stays constant over the same time interval. We can then write

$$\theta(t_{n+1}^-) = \theta(t_n^+) + \frac{d}{dt}\theta(t_n^+)\Delta t_n. \qquad (3.119)$$

Here, Δt_n is the interval between two consecutive kicks

$$\Delta t_n = L\frac{\bar{\omega}_{n+1}}{k_{n+1}c^2}. \qquad (3.120)$$

Using equation (3.115, 3.119), we can then obtain the following result:

$$\theta_{n+1} = \theta_n - \frac{\omega_0 L}{c}\frac{\sqrt{1+|w_{n+1}|}}{\sqrt{|w_{n+1}|}} + k_0Ls_{n+1} \qquad (3.121)$$

where s_{n+1} is the sign of w_{n+1}.

We see that equations (3.118, 3.121) define a discrete mapping on the phase plane (w, θ), which can be written as

$$w_{n+1} = w_n + A \sin \theta_n \tag{3.122}$$

$$\theta_{n+1} = \theta_n - B\sqrt{1 + |w_{n+1}|^{-1}} + \tilde{\theta} s_{n+1}. \tag{3.123}$$

This mapping depends on two parameters, A and B, determined by

$$A = aL, \quad B = \frac{\omega_0}{c} L. \tag{3.124}$$

The map (3.123) also shows a constant phase shift determined by $\tilde{\theta} = k_0 L$. For small values of the adimensional variable w, which corresponds to small values of the photon wavenumber, this mapping reduces to the $\hat{L}$-mapping introduced in [125], which shows an intermittency behaviour for $A \gg 1$.

This means that the photons can suffer intermittency acceleration when they interact with an electrostatic wavepacket in the limit of small photon wavenumbers (when they are close to the cut-off conditions).

3.5   Photon Fermi acceleration

We shall now discuss a different mechanism for photon acceleration which can occur inside an electromagnetic cavity with moving boundaries. This can be seen as the photon version of the well-known mechanism for cosmic ray acceleration first proposed by Fermi [29], where charged particles can gain energy by bouncing back and forth between two magnetic clouds.

This mechanism became extremely successful and it is dominantly used in the current models for cosmic ray acceleration in shocks. In its simplest and most popular versions, the charged particle can be described by two-dimensional discrete maps [59].

We will show in this section that a similar mechanism can be applied to photons, and that the photon motion can also be described by a discrete mapping [30].

In the case of photons, the magnetic clouds of the original model are replaced by mirrors or by plasma walls with a sharp density gradient. We can assume that one of the walls is fixed at $x = 0$ and the other oscillates around $x = L_0$ with a frequency ω'.

This can be described by a plasma frequency space–time variation of the form

$$\omega_p^2(x, t) = \omega_{p0}^2 \begin{cases} f(x - L(t)) & x > L(t) \\ 0 & L(t) \geq x > 0 \ . \\ f(-x) & x < 0 \end{cases} \tag{3.125}$$

The function $f(x)$ describes the plasma density profile of the two plasma walls (assuming that they have identical profiles), to be specified later. The

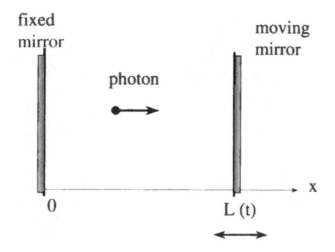

Figure 3.8. Fermi acceleration of photons inside an oscillating cavity.

moving wall is supposed to oscillate according to

$$L(t) = L_0(1 + \epsilon \cos \omega' t) \tag{3.126}$$

where the amplitude of this oscillation is assumed to be very small: $\epsilon \ll 1$.

The plasma walls are also supposed to be sufficiently dense in order to act as mirrors and to reflect all the incoming photons, which then remain trapped in a kind of one-dimensional cavity. Each time the photons are reflected by the moving plasma wall, they suffer a double Doppler shift, and their initial frequency ω_i is transformed into a final frequency ω_f, defined by the well-known law

$$\omega_f = \omega_i \frac{1 + \beta}{1 - \beta}. \tag{3.127}$$

The velocity of this moving wall is just the time derivative of $L(t)$ and we can write

$$\beta = -\epsilon \frac{L_0 \omega'}{c} \sin \omega' t. \tag{3.128}$$

If we define two adimensional quantities

$$b = \frac{L_0 \omega'}{c}, \quad \theta = \omega' t \tag{3.129}$$

we obtain the following law for the frequency shift after successive reflections at the moving plasma wall:

$$\omega_{n+1} = \omega_n \frac{1 - \epsilon b \sin \theta_n}{1 + \epsilon b \sin \theta_n}. \tag{3.130}$$

Now we have to find a corresponding transformation law for the phase θ between two successive reflections. Clearly, we can write $\theta_{n+1} = \theta_n + \omega' \Delta t$, where Δt is the time spent by the photon between two successive reflections at the moving wall. This time interval can be divided into two distinct parts: $\Delta t = \Delta t_0 + \Delta t_p$. The first part is the time spent by the photon in the vaccum region between the two plasma walls.

For $\epsilon \ll 1$, we can use, as a good approximation

$$\Delta t_0 = 2 \frac{L_0}{c}. \tag{3.131}$$

In order to calculate the time spent by the photon inside the two plasma regions, Δt_p, we notice that the photon velocity is determined by:

$$\frac{dx}{dt} = c \sqrt{1 - (\omega_{p0}/\omega)^2 f(x, t)}. \tag{3.132}$$

If the time spent inside the moving plasma region is much shorter than the period of the wall oscillations, $\Delta t_p \ll (4\pi/\omega')$, we can neglect the plasma motion during the process of photon reflection, and replace $f(x, t)$ by $f(x)$ in this equation. This leads to

$$\Delta t_p = \frac{4}{c} \int_0^{x_c} \frac{dx}{\sqrt{1 - (\omega_{p0}/\omega)^2 f(x)}}. \tag{3.133}$$

Here, the factor of 4 was introduced in order to account for the four distinct paths inside the two plasma walls. This integral extends from the plasma boundary to the cut-off position x_c, where the photon frequency equals the electron plasma frequency $\omega = \omega_p(x_c)$. Let us also introduce a normalized frequency

$$u = \frac{\omega}{\omega_{p0}}. \tag{3.134}$$

Using equations (3.130, 3.133), we can establish the following discrete map on the phase plane (u, θ):

$$u_{n+1} = u_n F(\theta_n) \tag{3.135}$$
$$\theta_{n+1} = \theta_n + G(u_{n+1}). \tag{3.136}$$

The function $F(\theta_n)$ is solely dependent on the double Doppler shift at the moving wall, and the function $G(u_{n+1})$ is determined by the plasma density profile

$$F(\theta) = \frac{1 - \epsilon b \sin \theta}{1 + \epsilon b \sin \theta} \tag{3.137}$$

$$G(u) = 2b \left(1 + \frac{2}{L_0} \int_0^{x_c} \frac{dx}{\sqrt{1 - f(x)/u^2}} \right). \tag{3.138}$$

Let us take the particularly interesting example of a parabolic density profile

$$f(x) = \frac{\alpha^2 \pi^2}{L_0^2} x^2 \tag{3.139}$$

where the parameter α defines the plasma density slope.

Then equation (3.138) reduces to

$$G(u) = 2b \left(1 + \frac{u}{\alpha} \right). \tag{3.140}$$

The map (3.135, 3.136) also takes a simpler form for very small plasma wall oscillations, such that $\epsilon b \ll 1$. In this case, we have

$$F(\theta) = 1 + 2\epsilon b \sin \theta. \tag{3.141}$$

We should note that all these maps are not area preserving. This can easily be seen by considering the Jacobian of the transformation $(u_n, \theta_n) \to (u_{n+1}, \theta_{n+1})$:

$$|J| = \begin{vmatrix} \partial u_{n+1}/\partial u_n & \partial u_{n+1}/\partial \theta_n \\ \partial \theta_{n+1}/\partial u_n & \partial \theta_{n+1}/\partial \theta_n \end{vmatrix} = |F(\theta)| \neq 1. \tag{3.142}$$

This is precisely the factor by which the photon frequency is double Doppler shifted by the moving plasma wall. However, because $F(\theta)$ is a periodic function of θ, we see that sets consisting of thin layers in phase space, of infinitesimal width δu and extending in phase from $\theta = 0$ to $\theta = 2\pi$, are area preserving.

It is also quite useful to determine the fixed points of the map, and their stability. This will give us important information concerning the qualitative aspects of photon motion. The first-order fixed points are determined by

$$u_{n+1} = u_n F(\theta_n) = u_n \tag{3.143}$$
$$\theta_{n+1} = \theta_n + G(u_{n+1}) = \theta_n + 2m\pi \tag{3.144}$$

for m integer. This is equivalent to

$$F(\theta) = 1, \quad G(u) = 2m\pi. \tag{3.145}$$

For the parabolic density profile (3.139), this leads to

$$\theta_m = 0, \quad u_m = \alpha \left(\frac{m\pi}{b} - 1 \right). \tag{3.146}$$

The stability of these fixed points can be determined by locally linearizing the map around each of them. Let us first linearize on the variable u, by defining

$$u = u_m + \bar{u}. \tag{3.147}$$

Replacing it in equations (3.135, 3.136, 3.141), we obtain

$$\bar{u}_{n+1} = \bar{u}_n + 2\epsilon u_m \sin \theta_n \tag{3.148}$$

$$\theta_{n+1} = \theta_n + \frac{2b}{\alpha}\bar{u}_{n+1} + 2m\pi. \tag{3.149}$$

Introducing a new variable I and a new parameter K, such that

$$I = \frac{2b}{\alpha}\bar{u}, \quad K = \frac{4b^2}{\alpha}\epsilon u_m, \tag{3.150}$$

this reduces to the well-known standard map, first studied by Chirikov [17, 59]:

$$I_{n+1} = I_n + K \sin \theta_n \tag{3.151}$$

$$\theta_{n+1} = \theta_n + I_{n+1}. \tag{3.152}$$

This shows that the photon dynamics around a given fixed point (θ_m, u_m) is approximately described by the standard map. It is well known that such a map suffers a topological transition into large-scale stochasticity if $K > 1$. This means that, near the fixed points, the photon motion will become stochastic in a significant fraction of the available phase space, if

$$u_m > \frac{\alpha}{4b^2\epsilon}. \tag{3.153}$$

We now examine the stability of the fixed points (3.145) of the Fermi mapping for the parabolic density profile. Following the usual procedure [59], we linearize the map on both variables, and obtain

$$\vec{x}_{n+1} = \mathbf{A} \cdot \vec{x}_n \tag{3.154}$$

where $\vec{x} = (\bar{u}, \bar{\theta}) = (u - u_m, \theta - \theta_m)$, and the matrix transformation $\mathbf{A}$ is the linearized Jacobian matrix

$$\mathbf{A} = \begin{bmatrix} 1 & -2\epsilon b u_m \\ \frac{2b}{\alpha} & 1 - 4\frac{\epsilon b^2}{\alpha}u_m \end{bmatrix}. \tag{3.155}$$

It can easily be seen that this linear map is area preserving:

$$|\mathbf{A}| = 1. \tag{3.156}$$

Stability of the fixed points (3.145) requires that $|\mathrm{Tr}\,\mathbf{A}| < 2$, or, more explicitly, that

$$u_m < \frac{\alpha}{\epsilon b^2}. \tag{3.157}$$

We can then say that, for high enough integers m such that this inequality is not verified, the first-order fixed points are all unstable. This means that, for such values of the photon normalized frequency u, the photon motion is essentially

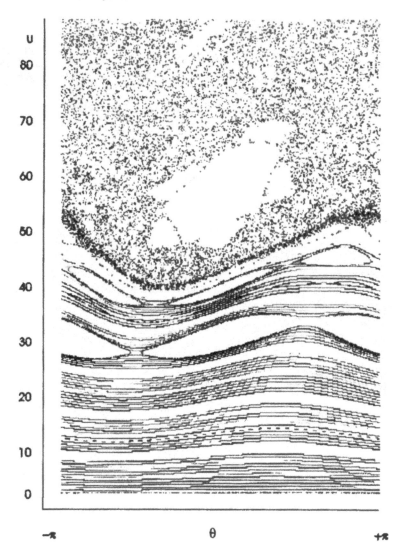

Figure 3.9. Phase space of the Fermi photon map. Regular and stochastic trajectories are shown.

stochastic. However, for much lower frequencies, significant stochastic motion had already taken place, according to the much less stringent threshold criterion (3.153). This is well illustrated by numerical calculations.

From here we conclude that a broad spectrum of radiation (which can be called white light) can be generated from nearly monochromatic light trapped inside an oscillating plasma cavity.

It is interesting to compare this qualitative aspect of the photon Fermi map with the usual Fermi mappings for charged particles [59] where, in contrast with the threshold criterion given by equation (3.157), an upper limit exists for particle energies, above which they cannot be accelerated. In some sense the photon Fermi map is similar to the inverted Fermi map for charged particles, where the energy axis is turned upside down.

This model of Fermi acceleration of photons could, in principle, be experimentally tested using an oscillating optical cavity, if the cavity parameters were conveniently chosen and if the photons could bounce back and forth several times inside the cavity, before escaping out, or before being absorbed by the walls.

It can also be useful in the context of astrophysics. For instance, the spectrum emitted by some very high-redshift radio galaxies reveals a strong asymmetry in the Lyman α line profile, indicating a clear blueshift [115]. This can be interpreted as a result of Fermi acceleration of the Lyman α photons by a moving shock [11].

Finally, we should notice that the concept of photon acceleration in a plasma can be extended to the acceleration of plasmons, or quanta of electron plasma waves, when they are trapped inside an unstable cavity. In this case [24], the plasmon Fermi acceleration will mainly lead to a change in the plasmon wavenumber because the plasmon frequency is always nearly equal to the electron plasma frequency.

3.6 Magnetoplasmas and other optical media

In this chapter we have only focused on processes occurring in isotropic plasmas because most of the published work on photon acceleration concerns this medium. However, our theoretical approach remains valid for other optical media.

Let us then conclude the chapter with some brief comments on the photon processes associated with moving perturbations of the refractive index in magnetized plasmas and in non-ionized optical media, such as a neutral gas, a glass or an optical fibre.

The photon dispersion relation in a magnetized plasma depends, not only on the electron plasma frequency ω_p, but also on the value and direction of the static magnetic field $\vec{B}_0$. We can represent it generally as

$$R(\omega, \vec{k}; \omega_p(\vec{r}, t), \omega_c(\vec{r}, t)) = 0 \tag{3.158}$$

where $\omega_c = e|B_0|/m$ is the electron cyclotron frequency.

This means that we can expect to obtain photon acceleration, not only by using moving electron density perturbations, such as ionization fronts and wakefields, but also if we excite similar forms of moving magnetic field perturbations. They can be produced by non-stationary currents applied to external coils, or by propagating low-frequency electromagnetic waves, such as Alfven waves.

The explicit expression of this dispersion relation is quite complicated, even in the limit of high-frequency waves for which the ion dispersion effects can be neglected, but we can still consider some simple examples.

First, if we assume photon propagation in a direction perpendicular to the static magnetic field $\vec{B}_0$, there are two distinct polarization states. One corresponds to the ordinary mode, with photons linearly polarized along $\vec{B}_0$, for which the dispersion relation is identical to that considered before for a non-magnetized plasma. The other corresponds to the extraordinary mode, with photons elliptically polarized in the plane perpendicular to $\vec{B}_0$. The dispersion relation for this new mode is

$$\omega^2 = k^2 c^2 + \omega_p^2 \frac{\omega^2 - \omega_p^2}{\omega^2 - \omega_{uh}^2} \tag{3.159}$$

where $\omega_{uh} = \sqrt{\omega_p^2 + \omega_c^2}$ is called the upper-hybrid frequency (the lower-hybrid one would only appear for low-frequency waves where the influence of the ion motion has to be retained).

We see that, for a photon frequency close to this resonance frequency, a small time change in the static magnetic field, or equivalently in the electron cyclotron frequency ω_c, will significantly alter the value of the refractive index and will lead to a frequency shift. In order to calculate this effect with the aid of the canonical equations for the photons we have to make use of the Hamiltonian function, valid for the extraordinary mode.

From the above dispersion relation, we can easily get

$$\omega(\vec{r}, \vec{k}, t) = \left[\frac{1}{2} (k^2 c^2 + \omega_p^2 + \omega_{uh}^2) \right.$$
$$\left. + \frac{1}{2} \sqrt{(k^2 c^2 + \omega_p^2 + \omega_{uh}^2)^2 - 4(k^2 c^2 \omega_{uh}^2 + \omega_p^4)} \right]^{1/2} \tag{3.160}$$

where both the electron plasma frequency and the upper-hybrid frequency are functions of $\vec{r}$ and t.

Similarly, for propagation parallel to the static magnetic field, we have two photon polarization states, with the corresponding dispersion relation

$$\omega^2 = k^2 c^2 + \omega_p^2 \frac{\omega}{\omega \pm \omega_c}. \tag{3.161}$$

The plus sign corresponds to the L-mode, which is left circularly polarized, and the minus sign corresponds to the R-mode, which is right circularly polarized. Because, by definition, we have $\omega_c > 0$, we can see that the L-mode has no resonances, while the R-mode is resonant for $\omega = \omega_c$. This is the so-called cyclotron resonance. This means that a small space–time change in the cyclotron frequency will lead to a significant change in the refractive index for the R-mode.

However, in dealing with photon motion very close to resonances, we have to take into account the wave absorption mechanisms, which will eventually reduce the efficiency of the expected frequency shift or photon acceleration processes. These absorption mechanisms are well known in plasma physics and will not be discussed here.

Let us now turn to non-ionized optical media. The typical form for the linear dispersion relation in these media is

$$\omega^2 = k^2 c^2 + \frac{\omega^2 \omega_f^2}{\omega^2 - \omega_0^2} \tag{3.162}$$

where ω_0 is the frequency of the nearest resonant transition between two quantum levels and ω_f plays the role of the plasma frequency. It depends on the density of atoms or molecules in the medium and on the value of the transition probabilities between these two quantum levels.

For $\omega \gg \omega_0$ this dispersion relation reduces to that of the non-magnetized plasma. However, for visible light propagating in the usual optical media, we have $\omega \ll \omega_0$, which explains why the refractive index is usually greater than one (in contrast with the plasma case where it is less than one).

Two ways can be foreseen to produce a space–time change in the refractive index and a subsequent photon acceleration in such media. The first one is to change the transition frequency ω_0. This can, for instance, be done with the aid of an externally applied electric field. The resulting Stark effect will lead to a detuning (quite often a splitting) of the atomic transition energy levels. We can then imagine a physical configuration where the optical medium is located inside an elongated capacitor.

By applying a sudden voltage signal at one extremity of this capacitor, the voltage signal will propagate along the capacitor arms and produce a moving transition between two different values of the refractive index of the optical medium [28]. The closer we are to the resonance condition, the stronger will be the change in the refractive index.

A more drastic change in the refractive index can be obtained by using electromagnetic induced transparency (EIT) [38]. This corresponds to exactly equating to zero the value of ω_f, by inhibiting transitions between the two quantum states with the help of an auxiliary light source. The photons of this auxiliary source couple one of these two energy states with a third one. In this way, the refractive index of a nearly resonant medium (which is opaque, because it completely absorbs the resonant photons) can be reduced exactly to the vacuum value, corresponding to a complete transparency.

It should be noticed that, in recent experiments with EIT produced by auxiliary short laser pulses, a large spectral broadening is observed [39], which can be interpreted as the result of photon acceleration.

These ways of changing the refractive index of a medium are essentially linear, but we can also imagine similar effects resulting from nonlinear mechanisms. For instance, it is known that a strong laser beam can produce a change in the refractive index, according to

$$n = n_0 + n_2 I(\vec{r}, t) \tag{3.163}$$

where n_0 is the linear refractive index, n_2 is the nonlinear refractive index (proportional to the nonlinear susceptibility of the medium) and $I(\vec{r}, t)$ is the intensity profile of the strong laser beam.

If photons having a different frequency and belonging to a probe beam co-propagate with the first one and cross the strong-beam boundary (because their group velocities are different), they will suffer a frequency shift described by the above canonical equations and similar to the one observed for the ionization front. The difference here is that the frequency shift will be negative for a photon crossing the front of the strong beam, and positive for a photon crossing the rear of the beam, because n_2 is positive, in contrast with the ionization front case which led to a decrease in the refractive index.

This effect is well understood in the frame of our single photon dynamical approach, but it is currently known as *induced phase modulation*. It is quite clear that the field phase is not an essential aspect of the problem, because the phase is completely absent from our photon canonical equations, and we still get a frequency shift.

The problem of the influence of the field phase in the so-called phase modulation effects is very interesting and will be discussed in detail later. Here we only would like to stress that the photon dynamical theory discussed in this chapter can equally well treat the cases of co-propagating and counter-propagating photons, in contrast with the usual theory of induced phase modulation which is only used for the co-propagation case [3].

Chapter 4

Photon kinetic theory

Until now, we have considered single particle trajectories, which (as we have shown) can account for several new and interesting features of electromagnetic wavepackets travelling in a non-stationary medium. However, this approach is not capable of describing the change in the internal structure of the wavepackets themselves, because such a structure is simply forgotten.

The easiest way to obtain a more detailed description of the wavepackets is to describe them, not as a single photon, but as an ensemble of photons with a given spectral and space–time distribution. The electromagnetic wave spectrum will then be seen as a gas of photons evolving in an optical medium. The kinetic equation describing the space–time evolution of such a gas is derived here.

First, we will use simple and intuitive arguments which are similar to those leading to the Klimontovich equation for charged particles in a gas. A second and more elaborated method will also be described, where the analogue of the Wigner function for the electromagnetic field is introduced.

These kinetic equations sometimes contain too much information for several problems, for which a simpler description of the photon distributions is required. For that reason it is also interesting to derive photon conservation equations which allow us to describe the electromagnetic field as a fluid of identical particles, and to consider only averaged photon properties.

This chapter will be completed by a detailed discussion of the phase-space representation of a short electromagnetic pulse. Examples of time evolution of short pulses with chirp and frequency shift are given. In particular, we present a model for the self-induced blueshift, which is produced when a strong laser pulse propagates along a neutral gas and produces an ionization front.

We conclude the chapter by considering the well-known effect of self-phase modulation and by showing that it can be described as a particular example of photon acceleration.

4.1   Klimontovich equation for photons

Let us first consider a single photon trajectory, as defined by the ray equations in their canonical form. The dynamical state of the photon (position and local frequency) can be determined by its location in the six-dimensional phase space $(\vec{r}, \vec{k})$.

We can also associate with this single photon a microscopic density distribution defined by the product of two Dirac δ functions

$$N_1(\vec{r}, \vec{k}, t) = \delta[\vec{r} - \vec{r}(t)]\delta[\vec{k} - \vec{k}(t)] \tag{4.1}$$

where $\vec{r}(t)$ and $\vec{k}(t)$ represent the actual photon trajectory, as determined by the solutions of the ray equations.

Of course, in order to accurately describe the evolution of a given electromagnetic pulse propagating in a medium, it is more convenient to consider, instead of a single photon trajectory (which gives no information on the detailed structure of the pulse), a large number of nearby photon trajectories. Let us then assume n different photon trajectories, and the associated density distribution

$$N(\vec{r}, \vec{k}, t) = \sum_{j=1}^{n} \delta[\vec{r} - \vec{r}_j(t)]\delta[\vec{k} - \vec{k}_j(t)]. \tag{4.2}$$

Integration in phase space clearly gives the total number of photons (or ray trajectories)

$$n = \int d\vec{r} \int d\vec{k}\, N(\vec{r}, \vec{k}, t). \tag{4.3}$$

If we want to establish an equation for the time evolution of the photon density distribution, we can take the partial time derivative of equation (4.2). We have

$$\frac{\partial}{\partial t} N(\vec{r}, \vec{k}, t) = \sum_{j=1}^{n} \left(\frac{d\vec{r}_j}{dt} \cdot \frac{\partial}{\partial \vec{r}_j} + \frac{d\vec{k}_j}{dt} \cdot \frac{\partial}{\partial \vec{k}_j} \right) \delta[\vec{r} - \vec{r}_j(t)]\delta[\vec{k} - \vec{k}_j(t)]. \tag{4.4}$$

Now, we can use the obvious relation $\partial/\partial x f(x - y) = -\partial/\partial y f(x - y)$, and transform this expression into

$$\frac{\partial}{\partial t} N(\vec{r}, \vec{k}, t) = \sum_{j=1}^{n} \left(\frac{d\vec{r}_j}{dt} \cdot \frac{\partial}{\partial \vec{r}} + \frac{d\vec{k}_j}{dt} \cdot \frac{\partial}{\partial \vec{k}} \right) \delta[\vec{r} - \vec{r}_j(t)]\delta[\vec{k} - \vec{k}_j(t)]. \tag{4.5}$$

Another simplification can be introduced by using the following property of the Dirac δ function: $x\delta(x - y) = y\delta(x - y)$. This allows us to replace the remaining $\vec{r}_j$ and $\vec{k}_j$ by $\vec{r}$ and $\vec{k}$ inside the differential operator acting on the δ functions.

The result is

$$\frac{d}{dt} N(\vec{r}, \vec{k}, t) = 0 \tag{4.6}$$

with

$$\frac{d}{dt} = \frac{\partial}{\partial t} + \frac{d\vec{r}}{dt} \cdot \frac{\partial}{\partial \vec{r}} + \frac{d\vec{k}}{dt} \cdot \frac{\partial}{\partial \vec{k}}. \tag{4.7}$$

Equation (4.6) can be called a kinetic equation for photons in its Klimontovich form [49]. It simply states that the photon number density is conserved. The problem with this simple result is that it was obtained by assuming that the photons are point particles, which is obviously not the case, and its validity has to be confirmed *a posteriori* by using Maxwell's equations. We will see below that this equation is only approximately valid, as should be expected, and that its range of validity is nearly (but not exactly) coincident with the range of validity of geometric optics.

From the canonical ray equations, we see that this kinetic equation can also be written as

$$\frac{\partial N_k}{\partial t} + [N_k, \omega] = 0 \tag{4.8}$$

where $N_k \equiv N_k(\vec{r}, t) \equiv N(\vec{r}, \vec{k}, t)$, and the Poisson bracket is

$$[N_k, \omega] = \frac{\partial N_k}{\partial \vec{r}} \cdot \frac{\partial \omega}{\partial \vec{k}} - \frac{\partial N_k}{\partial \vec{k}} \cdot \frac{\partial \omega}{\partial \vec{r}}. \tag{4.9}$$

At this point we could follow the usual statistical procedure and introduce some coarse-graining in the photon phase space, which would replace the quite spiky quantity $N_k(\vec{r}, t)$ by a smooth and well-behaved function like its ensemble average $\langle N_k(\vec{r}, t) \rangle$. This would lead us too far from our present purpose. It is more interesting here to establish a link between this quantity and the electromagnetic energy density.

By definition, we can write the total energy as

$$W(t) = \int w(\vec{r}, t) \, d\vec{r} \tag{4.10}$$

where the energy density is

$$w(\vec{r}, t) = 2 \int \hbar \omega_k N_k(\vec{r}, t) \frac{d\vec{k}}{(2\pi)^3}. \tag{4.11}$$

This equation states that the energy of each photon is equal to $\hbar \omega_k$, as we know from quantum theory. The factor of 2 is introduced because of the existence of two possible states of polarization. On the other hand, we can exactly establish, from the classical theory of radiation [57], that

$$w(\vec{r}, t) = \frac{\epsilon_0}{4} \int \left(\frac{\partial \omega R}{\partial \omega} \right)_k |E_k|^2 \frac{d\vec{k}}{(2\pi)^3} \tag{4.12}$$

where $|E_k|^2$ is the module square of the electric field amplitude of the Fourier component $\vec{k}$, and $R \equiv R(\omega, \vec{k}) = 0$ is the dispersion relation of the medium:

$$R(\omega, \vec{k}) = \epsilon(\omega, \vec{k}) - \frac{k^2 c^2}{\omega^2} + |\vec{k} \cdot \hat{e}|^2 \frac{c^2}{\omega^2} = 0. \tag{4.13}$$

Here $\hat{e}$ is the unit polarization vector. By comparing these two expressions for the electromagnetic energy density, we obtain

$$N_k(\vec{r}, t) = \frac{\epsilon_0}{8\hbar} \left(\frac{\partial R}{\partial \omega} \right)_k |E_k|^2. \tag{4.14}$$

Obviously, this definition of the photon number density can only make sense if we assume that the electric field amplitude of each Fourier component is a slowly varying function of space and time. A more refined way of establishing the definition of $N_k(\vec{r}, t)$ is based on the concept of the Wigner functions for the electromagnetic field, which will be considered next.

4.2   Wigner–Moyal equation for electromagnetic radiation

4.2.1   Non-dispersive medium

We will consider first a non-dispersive medium, in order to clearly state our procedure. We will also assume that the medium is isotropic and with no losses. In the absence of charge and current distributions, we have, from Maxwell's equations,

$$\nabla^2 \vec{E} - \nabla(\nabla \cdot \vec{E}) - \mu_0 \frac{\partial^2}{\partial t^2} \vec{D} = 0 \tag{4.15}$$

where $\vec{D} = \epsilon_0 \epsilon \vec{E}$ is the displacement vector. We also have $\epsilon = 1 + \chi$, where χ is the susceptibility of the medium.

Assuming, for simplicity, that the fields are transverse ($\nabla \cdot \vec{E} = 0$), we can write

$$\nabla^2 \vec{E} - \frac{1}{c^2} \frac{\partial^2 \vec{E}}{\partial t^2} = \frac{1}{c^2} \frac{\partial^2}{\partial t^2} (\chi \vec{E}). \tag{4.16}$$

For a wave with a given frequency ω and wavenumber $\vec{k}$, we can define the Wigner function for the electric field as

$$F(\vec{r}, t; \omega, \vec{k}) = \int d\vec{s} \int d\tau \vec{E} \left(\vec{r} + \frac{\vec{s}}{2}, t + \frac{\tau}{2} \right) \cdot \vec{E}^* \left(\vec{r} - \frac{\vec{s}}{2}, t - \frac{\tau}{2} \right) e^{-i\vec{k}\cdot\vec{s} + i\omega\tau}. \tag{4.17}$$

This quantity is formally quite similar to the Wigner function for a quantum system [40]. In contrast with our classical approach, the quantum Wigner function is well understood and of current use in quantum optics [58, 116].

Following a procedure explained in detail in appendix A, we can derive from the above wave equation an equation describing the space–time evolution of the Wigner function. The same procedure was used in reference [111] to study the case of relativistic plasmas.

In our case of a non-dispersive medium, the evolution equation for $F(\vec{r}, t; \omega, \vec{k})$ takes the form

$$\left(\epsilon \frac{\partial}{\partial t} + \frac{c^2 \vec{k}}{\omega} \cdot \nabla \right) F + \left(\frac{\partial \epsilon}{\partial t} \right) F = -\omega(\epsilon \sin \Lambda F) \tag{4.18}$$

where Λ is a differential operator, which acts both backwards on ϵ and forwards on F. It can be defined by

$$\Lambda = \frac{1}{2} \overset{\leftarrow}{} \left[\frac{\partial}{\partial \vec{r}} \cdot \frac{\partial}{\partial \vec{k}} - \frac{\partial}{\partial t} \frac{\partial}{\partial \omega} \right] \overset{\rightarrow}{}. \tag{4.19}$$

The right and left arrows are here to remind us that, in each of the two terms, the first differential operator acts backwards on ϵ and the second one acts forwards on F. The sine differential operator in equation (4.18) is, in fact, an infinite series of differential operators, according to

$$\sin \Lambda = \sum_{l=0}^{\infty} \frac{(-1)^l}{(2l+1)!} \Lambda^{2l+1}. \tag{4.20}$$

At the cost of such an unusual operator, we were able to derive from Maxwell's equations a closed evolution equation for the Wigner function F of the electric field. This is valid in quite general conditions, apart from our basic assumptions that the medium should be non-dispersive and that the dielectric constant should only evolve on a slow timescale. Its relation to the geometric optics approximation will become apparent below.

Equation (4.18) is formally quite similar to the Wigner–Moyal equation for quantum systems [40, 80], except for the term on the time derivative of the refractive index, which has no equivalent in the quantum mechanical problem. For that reason it can be called the Wigner–Moyal equation for the electromagnetic field. Clearly, it is significantly more complex than the kinetic equation established at the begining of this chapter.

In order to compare these two approaches, it is useful to introduce a few simplifying assumptions. The first one is associated with the character of the electromagnetic spectrum. We can assume that this spectrum is just a superposition of linear waves. For each spectral component, the value of the frequency ω has to satisfy the linear dispersion relation of the medium

$$\omega = \omega_k = kc/\sqrt{\epsilon}. \tag{4.21}$$

The corresponding group velocity is

$$\vec{v}_k = \frac{\partial \omega_k}{\partial \vec{k}} = \frac{c}{\sqrt{\epsilon}} \frac{\vec{k}}{k} = \frac{c^2}{\omega_k \epsilon} \vec{k}. \tag{4.22}$$

In this case of a linear wave spectrum, the Wigner function F takes the form

$$F \equiv F(\vec{r}, t; \omega, \vec{k}) = F_k(\vec{r}, t)\delta(\omega - \omega_k). \tag{4.23}$$

Replacing it in equation (4.26), and noting that the reduced Wigner function F_k is independent of ω and consequently that

$$\frac{\partial^m F}{\partial \omega^m} = F_k \frac{\partial^m}{\partial \omega^m} \delta(\omega - \omega_k) = (-1)^m \delta(\omega - \omega_k) \frac{\partial^m F_k}{\partial \omega^m} = 0, \tag{4.24}$$

we can write the Wigner–Moyal equation in a simplified form

$$\left(\frac{\partial}{\partial t} + \vec{v}_k \cdot \nabla\right) F_k + \frac{\partial \ln \epsilon}{\partial t} F_k = -\frac{\omega_k}{\epsilon} \left[\epsilon \sin \Lambda_k F_k\right]. \qquad (4.25)$$

Here, Λ_k is a reduced differential operator defined by

$$\Lambda_k = \frac{1}{2} \frac{\overleftarrow{\partial}}{\partial \vec{r}} \cdot \frac{\overrightarrow{\partial}}{\partial \vec{k}} . \qquad (4.26)$$

Because, in the Wigner–Moyal equation, the sine operators are too complicated to be calculated in specific problems, it is useful to simply retain the first term in the development (4.20):

$$\sin \Lambda_k \simeq \Lambda_k. \qquad (4.27)$$

This is valid for a slowly varying medium, where the gradients contained in the operator Λ_k are very small. In such a case, we are close to the conditions where the geometric optics approximation is valid, and the Wigner–Moyal equation reduces to

$$\left(\frac{\partial}{\partial t} + \vec{v}_k \cdot \nabla\right) F_k + \left(\frac{\partial \ln \epsilon}{\partial t}\right) F_k \simeq -\frac{\omega_k}{2\epsilon} \left(\frac{\partial \epsilon}{\partial \vec{r}} \cdot \frac{\partial F_k}{\partial \vec{k}}\right). \qquad (4.28)$$

On the other hand, if we neglect the logarithmic derivative in this equation, we notice that this equation implies that a triple equality exists, namely

$$dt = \frac{d\vec{r}}{\vec{v}_k} = \frac{d\vec{k}}{(\omega_k/2\epsilon)(\partial \epsilon/\partial \vec{r})}. \qquad (4.29)$$

This is equivalent to stating that

$$\frac{d\vec{r}}{dt} = \vec{v}_k = \frac{\partial \omega_k}{\partial \vec{k}} \qquad (4.30)$$

$$\frac{d\vec{k}}{dt} = \frac{\omega_k}{2\epsilon} \frac{\partial \epsilon}{\partial \vec{r}} = \frac{kc}{2\epsilon^{3/2}} \frac{\partial \epsilon}{\partial \vec{r}} = -\frac{\partial \omega_k}{\partial \vec{r}}. \qquad (4.31)$$

We recover here the ray equations of the geometric optics approximation, identical to those used before. They are nothing but the characteristic equations of the simplified version of the Wigner–Moyal equation, which can then be written as

$$\frac{d}{dt} F_k \equiv \left(\frac{\partial}{\partial t} + \vec{v}_k \cdot \nabla + \frac{d\vec{k}}{dt} \cdot \frac{\partial}{\partial \vec{k}}\right) F_k \simeq 0. \qquad (4.32)$$

This equation, which states the conservation of the Wigner function F_k, is valid when the logarithmic time derivative, as well as the higher order derivatives associated with the diffraction terms $l > 0$ in the development of the sine operator $\sin \Lambda_k$, can be neglected. Furthermore, from equation (4.17) we can define $F_k(\vec{r}, t)$ as the space Wigner function for the electric field, as shown in appendix A:

$$F_k(\vec{r}, t) = \int \vec{E}(\vec{r} + \vec{s}/2, t) \cdot \vec{E}^*(\vec{r} - \vec{s}/2, t) e^{-i\vec{k} \cdot \vec{s}} \, d\vec{s}. \tag{4.33}$$

It is now useful to define the number of photons $N_k(\vec{r}, t)$ in terms of this reduced Wigner function, as

$$N_k(\vec{r}, t) = \frac{\epsilon_0}{8\hbar} \left(\frac{\partial R}{\partial \omega} \right)_{\omega_k} F_k(\vec{r}, t) \tag{4.34}$$

where $R = 0$ is the dispersion relation of the medium. For the case considered here of a non-dispersive medium it reduces to

$$R \equiv R(\omega, \vec{k}) = \epsilon - c^2 k^2 / \omega^2 = 0. \tag{4.35}$$

The expression for the number of photons (4.34) is then reduced to

$$N_k(\vec{r}, t) = \frac{\epsilon_0}{4\hbar} \frac{\epsilon}{\omega_k} F_k(\vec{r}, t). \tag{4.36}$$

We can now return to the somewhat more exact expression for the Wigner–Moyal equation (4.28) and rewrite it as

$$\frac{d}{dt} F_k = - \left(\frac{\partial \ln \epsilon}{\partial t} \right) F_k \tag{4.37}$$

where the total derivative is determined by equation (4.32).

On the other hand, if we take the total time derivative of the number of photons (4.36), and if we notice that

$$\frac{d\omega_k}{dt} = \frac{\partial \omega_k}{\partial t} = -\frac{\omega_k}{2} \left(\frac{\partial \ln \epsilon}{\partial t} \right), \tag{4.38}$$

we can then obtain

$$\frac{d}{dt} N_k = \left[\left(\frac{1}{2} \frac{\partial}{\partial t} + \vec{v}_k \cdot \nabla \right) \ln \epsilon \right] N_k. \tag{4.39}$$

Neglecting the slow variations of the refractive index appearing on the right-hand side, we can finally state an equation of conservation for the number of photons, in the form

$$\frac{dN_k}{dt} \equiv \left(\frac{\partial}{\partial t} + \vec{v}_k \cdot \nabla + \frac{d\vec{k}}{dt} \cdot \frac{\partial}{\partial \vec{k}} \right) N_k = 0. \tag{4.40}$$

This is identical to the Klimontovich equation derived at the begining of this chapter. This new derivation, which is much more complicated, has however the advantage of using a more general definition for N_k.

On the other hand, we understand from this that the conservation equation for the number of photons is only valid when the higher order terms contained in the sine operator of the Wigner–Moyal equation can be neglected. This means that these terms represent the diffraction corrections to the geometric optics approximation.

Finally, we note that the classical Wigner function for the electromagnetic field introduced in this section is sometimes used to characterize ultra-short laser pulses with a time-dependent spectrum, as measured in optical experiments [45]. In contrast, an evolution equation of this quantity seems to have been ignored. The Wigner–Moyal equation described here can eventually be used to understand the space–time evolution of such short pulses along a given optical circuit.

4.2.2   Dispersive medium

The above derivation is conceptually quite interesting because it establishes a clear link between the exact Maxwell's equations and the heuristically derived Klimontovich equation. However, its range of validity is not very wide because we have neglected dispersion effects.

The generalization to the case of a dispersive medium is considered in this section. For simplicity, we still neglect the losses in the medium, which can easily be included afterwards.

First of all, if the electromagnetic radiation propagates in a dispersive medium, our starting equation (4.16) has to be replaced by

$$\left(\nabla^2 - \frac{1}{c^2}\frac{\partial^2}{\partial t^2}\right)\vec{E} = \mu_0 \frac{\partial^2}{\partial t^2}\vec{P} \tag{4.41}$$

where $\vec{P} = \epsilon_0 \vec{E} - \vec{D}$ is the vector polarization of the medium. In general terms it can be related to the electric field $\vec{E}$ by the integral

$$\vec{P}(\vec{r},t) = \epsilon_0 \int d\vec{r}' \int dt' \chi(\vec{r},t,\vec{r}',t')\vec{E}(\vec{r}-\vec{r}',t-t'). \tag{4.42}$$

Again, we can derive from here an evolution equation for the double Wigner function for the electric field. The derivation is detailed in appendix B and, to the lowest order of the space and time variations of the medium, the result is

$$\left(\frac{\partial}{\partial t} + \vec{v}_g \cdot \nabla\right)F = -\frac{2}{2\omega + \partial\eta_0/\partial\omega}\,(\eta \wedge F) \tag{4.43}$$

where $\wedge$ is the differential operator defined by equation (4.19) and $\vec{v}_g$ is the group velocity defined by

$$\vec{v}_g = \frac{2c^2\vec{k} - \omega^2\partial\epsilon/\partial\vec{k}}{2\omega\epsilon + \omega^2\partial\epsilon/\partial\omega} \tag{4.44}$$

with $\epsilon = 1 + \chi = 1 + \eta/\omega^2$.

This means that, if we had used the full developement of $\eta_\pm$ around $(\vec{r}, t)$, instead of the first terms, we would have obtained the operator $\sin \Lambda$ instead of just Λ, characteristic of the Wigner–Moyal equation. This equation therefore generalizes the above derivation of this equation to the case of a dispersive medium. Obviously, for a non-dispersive medium, such that $\partial \eta_0/\partial \omega = 0$, this would reduce to the result of the previous section.

Let us assume that the electromagnetic wave spectrum is made of a superposition of linear waves, such that we can use equation (4.23), $F = F_k \delta(\omega - \omega_k)$. Then, equation (4.43) becomes

$$\left(\frac{\partial}{\partial t} + \vec{v}_k \cdot \frac{\partial}{\partial \vec{k}} + \frac{1}{(\partial \omega^2 \epsilon/\partial \omega)_{\omega_k}} \frac{\partial \eta_k}{\partial \vec{r}} \cdot \frac{\partial}{\partial \vec{k}} \right) F_k = 0 \qquad (4.45)$$

where $\vec{v}_k = (\partial \omega/\partial \vec{k})_{\omega_k}$ and $\eta_k = \omega_k^2 \chi(\vec{r}, \vec{k}, t)$.

As an example of a dispersive medium, we can consider an isotropic plasma, where we have $\eta_k = -\omega_p^2$. In this case, the gradient of η_k appearing in the last term of this equation reduces to the gradient of the electron plasma density, or equivalently, to the gradient of the square of the plasma frequency.

We have then

$$\left(\frac{\partial}{\partial t} + \vec{v}_k \cdot \frac{\partial}{\partial \vec{k}} - \frac{1}{2\omega_k} \frac{\partial \omega_p^2}{\partial \vec{r}} \cdot \frac{\partial}{\partial \vec{k}} \right) F_k = 0 \qquad (4.46)$$

where $\omega_k = \sqrt{k^2 c^2 + \omega_p^2(\vec{r}, t)}$.

This is equivalent to stating that the reduced Wigner function F_k is conserved:

$$\frac{d}{dt} F_k \equiv \left(\frac{\partial}{\partial t} + \vec{v}_k \cdot \frac{\partial}{\partial \vec{r}} + \frac{d\vec{k}}{dt} \cdot \frac{\partial}{\partial \vec{k}} \right) F_k = 0 \qquad (4.47)$$

because we know, from the photon ray equations, that

$$\frac{d\vec{k}}{dt} = -\frac{\partial \omega_k}{\partial \vec{r}} = -\frac{1}{2\omega_k} \frac{\partial \omega_p^2}{\partial \vec{r}}. \qquad (4.48)$$

From these reduced forms of the Wigner–Moyal equation for a dispersive medium, we can then justify the use of the equation of conservation for the number of photons $N_k(\vec{r}, t)$, equation (4.40), which can also be called the kinetic equation for photons propagating in slowly varying dispersive media. The practical interest of this kinetic approach will now be illustrated with a few specific examples.

4.3 Photon distributions

In order to illustrate the interest of this kinetic approach, let us give some examples and introduce some definitions. First of all, it should be noticed that we have introduced in equation (4.34) a more general definition for the number of photons than those usually found in the literature. As it states, it can be applied to arbitrary forms of wavefields (plane, spherical or cylindrical waves).

In particular, if we take the simple case of plane waves, such that $\vec{E}(\vec{r}, t) = \vec{E}_0 \exp(i\vec{k}_0 \cdot \vec{r} - i\omega_0 t)$, our definition reduces to

$$N_k(\vec{r}, t) = \frac{\epsilon_0}{8\hbar} \frac{\partial R}{\partial \omega} |E_0|^2 \delta(\vec{k} - \vec{k}_0). \tag{4.49}$$

This is just the definition commonly found in the literature [93, 114] which is not very useful to describe, for instance, short laser pulses.

4.3.1 Uniform and non-dispersive medium

Let us now consider some examples of solutions of the photon kinetic equation. For simplicity, we will discuss one-dimensional propagation. The pertinent kinetic equation will be

$$\left(\frac{\partial}{\partial t} + v_k \frac{\partial}{\partial x} + f_k \frac{\partial}{\partial k}\right) N_k(x, t) = 0. \tag{4.50}$$

The simplest possible case corresponds to a photon beam propagation in a uniform and non-dispersive medium. The third term in this equation will then be equal to zero, due to uniformity:

$$f_k \equiv \frac{dk}{dt} = -\frac{\partial \omega}{\partial x} = \frac{kc}{n^2} \frac{\partial n}{\partial x} = 0 \tag{4.51}$$

where the refractive index n is time independent.

Furthermore, in a non-dispersive medium, we also have $v_k = v_0 = c/n = $ const. This means that the one-dimensional kinetic equation (4.50) is reduced to

$$\left(\frac{\partial}{\partial t} + v_0 \frac{\partial}{\partial x}\right) N_k(x, t) = 0. \tag{4.52}$$

A possible solution of this equation is a Gaussian pulse, with a duration τ, which propagates along the medium without changing its shape. This can be represented by

$$N_k(x, t) = N(k) \exp\left[-(x - v_0 t)^2 / \sigma_x^2\right] \tag{4.53}$$

where $\sigma_x = v_0 \tau$ is the spatial pulse width, and $N(k)$ describes the spectral content.

Taking its space and time derivatives, we can easily see that this solution satisfies equation (4.52) for an arbitrary function $N(k)$. A useful choice is that of a spectral Gaussian distribution, centred around some wavenumber value k_0, with a spectral width σ_k

$$N(k) = N_0 \exp\left[-(k - k_0)^2 / \sigma_k^2\right]. \tag{4.54}$$

Compatibility between the spectral and the spatial distributions implies that we always have $\sigma_k \geq 1/\sigma_x$. The equality $\sigma_k \sigma_x = 1$ corresponds to the case of a transform limited pulse, which minimizes the position–momentum uncertainty relations.

4.3.2 Uniform and dispersive medium

As a first step in complexity, we consider a uniform but dispersive medium, where the refractive index is frequency dependent. The third term in the photon kinetic equation (4.50) will still be equal to zero, but the group velocity v_k will not be a constant. In order to calculate its value, let us linearize the refractive index around the central pulse frequency

$$n(\omega) \simeq n_0 + (\omega - \omega_0)\left(\frac{dn}{d\omega}\right)_{\omega_0} \tag{4.55}$$

where $\omega_0 = \omega(k_0)$ and $n_0 = n(\omega_0)$. The dispersion relation can then be written as

$$kc = n_0\omega + (\omega - \omega_0)\omega n_0' \tag{4.56}$$

with $n_0' = (dn/d\omega)_{\omega_0}$. This can be solved for ω to give

$$\omega = \frac{1}{2}\left(\omega_0 - \frac{n_0}{n_0'}\right) + \frac{1}{2}\sqrt{(\omega_0 - n_0/n_0')^2 + 4kc/n_0'}. \tag{4.57}$$

The group velocity becomes

$$v_k = \frac{\partial\omega}{\partial k} = \frac{c}{\sqrt{b^2 + 4kcn_0'}} \tag{4.58}$$

with $b = (\omega_0 n_0' - n_0)$.

We see that, for $n_0' > 0$, the higher frequencies inside the pulse (with higher values of k) will travel with lower velocities and will be retarded along propagation $v_k < v_0$, for $k > k_0$. In such a medium, the pertinent equation to be solved is

$$\left(\frac{\partial}{\partial t} + v_k \frac{\partial}{\partial x}\right) N_k(x, t) = 0. \tag{4.59}$$

Instead of the Gaussian distribution (4.53), we can try here a solution of the form

$$N_k(x, t) = N(k) \exp\left[-(x - v_k t)^2/\sigma_x^2\right] \tag{4.60}$$

wher v_k is determined by equation (4.58) and is independent of x and t.

This is clearly a solution of equation (4.59). Let us see the meaning of this new solution and compare it with (4.53), by linearizing v_k around k_0, $v_k \simeq v_0 + (k - k_0)v_0'$, with

$$v_0' = \left(\frac{\partial v_k}{\partial k}\right)_{k_0} = -\frac{2c^2 n_0'}{(b^2 + 4k_0 c n_0')^{3/2}}. \tag{4.61}$$

Replacing this development of v_k in equation (4.60), and using $N(k)$ as given by equation (4.54), we obtain a photon distribution formally analogous to (4.53), but with $N(k)$ replaced by a space- and time-dependent function:

$$N_k(x, t) = N(k, x, t) \exp\left[-(x - v_0 t)^2/\sigma_x^2\right] \qquad (4.62)$$

with

$$N(k, x, t) = N_0 \exp\left[-(k - k_0)^2/\sigma_k(t)^2\right]$$
$$\times \exp\left[2t(k - k_0)v_0'(x - v_0 t)/\sigma_x^2\right]. \qquad (4.63)$$

Here, we have used

$$\frac{1}{\sigma_k(t)^2} = \frac{1}{\sigma_k^2} + \frac{v_0'^2 t^2}{\sigma_x^2}. \qquad (4.64)$$

4.3.3 Pulse chirp

The above result shows that, when an initially Gaussian beam propagates in a uniform but dispersive medium it maintains its spatial Gaussian shape but its spectral distribution is distorted in time. At this point, the concept of pulse chirp has to be introduced, because it is associated with such a pulse distortion.

In the frame of our kinetic description of a photon beam, the chirp is determined by the space–time distribution of the averaged wavenumber:

$$\langle k \rangle = \frac{1}{n_\gamma(x, t)} \int k N_k(x, t) \frac{dk}{2\pi} \qquad (4.65)$$

where the normalization factor

$$n_\gamma = \int N_k(x, t) \frac{dk}{2\pi} \qquad (4.66)$$

is the photon mean density.

The concept of pulse chirp is very important for the optics of short laser pulses. We can say that a given pulse (or photon beam) is chirped if its averaged wavenumber is not constant across the pulse. As an example, let us consider the distribution (4.62–4.64). Replacing it in equation (4.65), we obtain

$$\langle k \rangle = \frac{\int_{-\infty}^{\infty} k \, dk \exp(-ak^2 + 2bk)}{\int_{-\infty}^{\infty} dk \exp(-ak^2 + 2bk)} \qquad (4.67)$$

where we have used

$$a = \frac{1}{\sigma_k(t)^2}, \quad b = \frac{k_0}{\sigma_k(t)^2} + \frac{v_0'}{\sigma_x^2}(x - v_0 t)t. \qquad (4.68)$$

Using the following solutions for these two integrals [35]:

$$\int_{-\infty}^{\infty} \exp(-ak^2 + 2bk)\, dk = \sqrt{\frac{\pi}{a}} \exp\left(\frac{b^2}{a}\right) \tag{4.69}$$

and

$$\int_{-\infty}^{\infty} k \exp(-ak^2 + 2bk)\, dk = \frac{b}{a}\sqrt{\frac{\pi}{a}} \exp\left(\frac{b^2}{a}\right), \tag{4.70}$$

we obtain

$$\langle k \rangle = \frac{b}{a} = k_0 + v_0' \frac{\sigma_k(t)^2}{\sigma_x^2}(x - v_0 t)t. \tag{4.71}$$

This shows that the mean value of the wavenumber inside the photon beam is time dependent, which means that propagation of a Gaussian pulse in a dispersive medium leads to pulse chirping, simply because some of the photons will travel with larger velocities than the others. In a similar way, we can calculate the spectral width of the chirp, by defining a new mean value

$$\langle k^2 \rangle = \frac{1}{n_\gamma(x, t)} \int k^2 n_k(x, t)\frac{dk}{2\pi}. \tag{4.72}$$

In our case, it can be written as

$$\langle k^2 \rangle = \frac{\int_{-\infty}^{\infty} k^2\, dk \exp(-ak^2 + 2bk)}{\int_{-\infty}^{\infty} dk \exp(-ak^2 + 2bk)}. \tag{4.73}$$

Using the following solution for the integral in the numerator [35]:

$$\int_{-\infty}^{\infty} k^2 \exp(-ak^2 + 2bk)\, dk = \frac{1}{2a}\sqrt{\frac{\pi}{a}}\left(1 + \frac{2b^2}{a}\right)\exp\left(\frac{b^2}{a}\right), \tag{4.74}$$

we obtain

$$\langle k^2 \rangle = \frac{1}{2a}\left(1 + \frac{2b^2}{a}\right) = \frac{\sigma_k(t)^2}{2} + \langle k \rangle^2. \tag{4.75}$$

This is equivalent to writing

$$\langle (\Delta k)^2 \rangle \equiv \langle (k - \langle k \rangle)^2 \rangle = \frac{\sigma_k(t)^2}{2}. \tag{4.76}$$

According to the definition of $\sigma_k(t)$, as given by equation (4.64), this means that the square mean deviation decreases with time, while the pulse chirp associated with the mean value $\langle k \rangle$ increases in time. Starting from a Gaussian spectral distribution, at $t = 0$, such that $\langle k \rangle$ is constant across the pulse spatial width, we obtain a thin and elongated spectral distribution as time evolves (or as the pulse propagates).

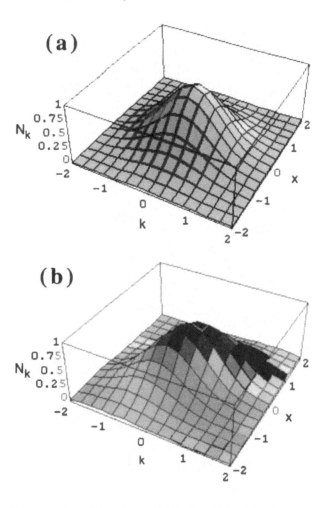

Figure 4.1. Representation of the photon distribution $N_k(x, t)$ in phase space (x, k), at a given time t, for (a) a transform limited Gaussian pulse; (b) a chirped pulse.

A more general class of solutions for the photon kinetic equation for uniform dispersive media can be stated as

$$N_k(x, t) = N(k, x, t) \exp\left[-(x - v_k t)^2/\sigma_x^2\right] \tag{4.77}$$

with

$$N(k, x, t) = N_0 \, e^{-[k - g(x,t)]^2/\sigma_k^2}. \tag{4.78}$$

Using this in equation (4.59) we can easily realize that this is a solution of the photon kinetic equation, for a uniform but dispersive medium, for every function satisfying the equation

$$\left(\frac{\partial}{\partial t} + v_k \frac{\partial}{\partial x}\right) g = 0. \tag{4.79}$$

As an example of a solution, we can consider the family of functions

$$g(x, t) = k_0 + g_0(x - v_k t)^n \tag{4.80}$$

where k_0 and g_0 are constants and n is an integer.

4.3.4   Non-stationary medium

Let us now assume a non-dispersive but non-stationary and non-uniform medium, such that the refractive index is determined by

$$n \equiv n(x, t) = n_0[1 + \delta f(x, t)] \tag{4.81}$$

where $f(x, t)$ describes a small space–time perturbation and $\delta \ll 1$ is the scale of the perturbation.

The dispersion relation can then be written as

$$\omega = \frac{kc}{n_0[1 + \delta f(x, t)]} \simeq \frac{kc}{n_0}[1 - \delta f(x, t)] \tag{4.82}$$

and the photon, or group velocity, as

$$v_k \equiv v_g(x, t) \simeq v_0[1 - \delta f(x, t)] \tag{4.83}$$

where we have used $v_0 = c/n_0$.

The photon kinetic equation becomes

$$\left(\frac{\partial}{\partial t} + v_g(x, t)\frac{\partial}{\partial x} + k v_0 \delta \frac{\partial f}{\partial x}\frac{\partial}{\partial k}\right) N_k(x, t) = 0 \tag{4.84}$$

where we have used $v_0 = c/n_0$. Let us try a solution of the form (4.77, 4.78), but with v_k replaced by the constant v_0. Using this in equation (4.84) we notice that the function $g(x, t)$ has to satisfy the following equation:

$$\left(\frac{\partial}{\partial t} + v_g \frac{\partial}{\partial x}\right) g = \delta v_0 k \frac{\partial f}{\partial x} + \delta f v_0^2 \frac{\sigma_k^2}{\sigma_x^2} \frac{(x - v_0 t)}{(k - g)}. \tag{4.85}$$

For a given perturbation of the refractive index of the medium $f(x, t)$, this equation allows us to obtain the solution $g(x, t)$. However, such solutions, even if they exist, are in general very difficult to find.

In order to obtain an approximate solution for $g(x, t)$, we can assume that we are interested in the main region of the pulse, such that $x \simeq v_0 t$, and the last term in this equation can be neglected. If we also take $v_g \simeq v_0$, the equation for $g(x, t)$ is reduced to

$$\left(\frac{\partial}{\partial t} + v_0 \frac{\partial}{\partial x}\right) g \simeq \delta v_0 k \frac{\partial f}{\partial x}. \tag{4.86}$$

An important example is that of a wakefield perturbation of the refractive index, such that $f(x, t) = \cos k_p(x - ut)$. In this case, the approximate solution for $g(x, t)$ is

$$g(x, t) = k_0 + g_0 \cos k_p(x - ut) \tag{4.87}$$

with

$$g_0 = -\delta \frac{k v_0}{u - v_0}. \tag{4.88}$$

This shows that the spectral shifts are more important when the phase velocity of the perturbation is nearly equal to the group velocity of the photons, $u \sim v_0$, as already found in the analysis of single photon trajectories. But, of course, these analytical results are quite rough and, for this and other situations, numerical solutions of the photon kinetic equation are required.

4.3.5 Self-blueshift

An interesting result [91] was obtained from the numerical integration of the photon kinetic equation, for the self-blueshift of a laser pulse penetrating in a neutral gas. This blueshift is due to the sudden ionization of the gas and subsequent creation of an ionization front.

Self-blueshift is a particular aspect of photon acceleration where the space–time changes in the refractive index are due to the incident laser pulse itself. Experimentally, this has been well known since the early sixties [120, 122], and it was studied theoretically by several authors [41, 48].

Let us describe the ionization model. For very intense incident laser fields, the photoionization processes are described by the tunnelling ionization theory. If the initial density of the neutral atoms of the gas is $n_0(\vec{r}, t)$, this number will decrease due to field ionization as

$$\frac{dn_0(\vec{r}, t)}{dt} = -w_1(\vec{r}, t)n_0(\vec{r}, t) \tag{4.89}$$

where w_1 is the ionization rate for the atom.

The number of singly ionized atoms $n_1(\vec{r}, t)$ will then be determined by the balance equation

$$\frac{dn_1(\vec{r}, t)}{dt} = w_1(\vec{r}, t)n_0(\vec{r}, t) - w_2(\vec{r}, t)n_1(\vec{r}, t) \tag{4.90}$$

where w_2 is the probability of double ionization occurring.

In more general terms, the number density of j-charged ions is determined by

$$\frac{dn_j(\vec{r}, t)}{dt} = w_j(\vec{r}, t)n_{j-1}(\vec{r}, t) - w_{j+1}(\vec{r}, t)n_j(\vec{r}, t). \qquad (4.91)$$

Finally, the density of completely ionized atoms is given by

$$\frac{dn_Z(\vec{r}, t)}{dt} = w_z(\vec{r}, t)n_Z(\vec{r}, t) \qquad (4.92)$$

where Z is the atomic number of the atoms in the gas.

The various ionization probabilities w_j depend on the local amplitude of the laser electric field and are well known from the standard theory of field ionization [4, 47]. The values of these ionization probabilities can establish the coupling between these balance equations for the atomic states and the above kinetic equation for the photons. Furthermore, the local plasma dispersion relation can be consistently determined by considering charge neutrality.

The electron plasma density will then be determined by

$$n_e(\vec{r}, t) = \sum_{j=0}^{j=Z} jn_j(\vec{r}, t). \qquad (4.93)$$

A numerical integration of the kinetic equation for photons, coupled with these simple balance equations, clearly shows that the front of a Gaussian beam penetrating in a neutral gas region will be significantly blueshifted (see figure 4.2). Furthermore, the blueshift is not uniform and several wavelike perturbations can be observed in the up-shifted spectrum, one for each ionization state. These spectral undulations will be eventually attenuated by electron impact ionization which was not contained in this model.

4.4   Photon fluid equations

In this chapter we have been able to show that the photon kinetic equation can be derived from Maxwell's equations, in the limit of slowly space- and time-varying media. Even if the range of validity of the photon kinetic theory is limited, it nevertheless contains detailed information about the photon spectrum. It is then useful to derive from this equation a set of conservation laws describing the space–time behaviour of photon averaged quantities, in the same way as fluid equations can be derived from kinetic equations for the particles of an ordinary gas.

Let us first define the mean density of the photon gas as the integral of the photon distribution over the entire spectrum:

$$n_\gamma(\vec{r}, t) = 2 \int N_k(\vec{r}, t) \frac{d\vec{k}}{(2\pi)^3}. \qquad (4.94)$$

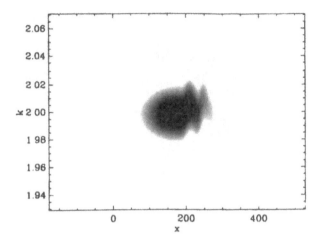

Figure 4.2. Numerical simulation of the self-blueshift of a laser beam with an intensity of 10^{15} W cm^{-2}, propagating in neutral argon, with an initial Gaussian spectrum. The initial density of the neutral atoms is 10^{18} cm^{-3} for $x > 0$.

Similarly, we can define the mean photon velocity, or mean group velocity, as

$$\vec{u}(\vec{r}, t) = \frac{2}{n_\gamma} \int \vec{v}_k N_k(\vec{r}, t) \frac{d\vec{k}}{(2\pi)^3}. \tag{4.95}$$

The factor of two appearing in these two definitions is due to the existence of the two independent polarization states. The continuity equation for the photon gas, or photon density conservation equation, can be derived by integrating the kinetic equation in $\vec{k}$. Integration of the first term gives

$$\frac{\partial}{\partial t} \int N_k \frac{d\vec{k}}{(2\pi)^3} = \frac{1}{2} \frac{\partial n_\gamma}{\partial t}. \tag{4.96}$$

Integration of the second term leads to

$$\int \vec{v}_k \cdot \nabla N_k \frac{d\vec{k}}{(2\pi)^3} = \nabla \cdot (\vec{u} n_\gamma) - \int N_k (\nabla \cdot \vec{v}_k) \frac{d\vec{k}}{(2\pi)^3}. \tag{4.97}$$

Finally, integration of the third term gives

$$\int \vec{f} \cdot \frac{\partial N_k}{\partial \vec{k}} \frac{d\vec{k}}{(2\pi)^3} = \int N_k \nabla \cdot \vec{v}_k \frac{d\vec{k}}{(2\pi)^3}. \tag{4.98}$$

The sum of these three contributions leads to the continuity equation for the photon gas

$$\frac{\partial n_\gamma}{\partial t} + \nabla \cdot (\vec{u} n_\gamma) = 0. \tag{4.99}$$

The second conservation equation to be derived is the photon momentum conservation law. Multiplying the photon kinetic equation by $\vec{v}_k$ and integrating it over the entire spectrum, we get the following contribution from the first term:

$$\int \vec{v}_k \frac{\partial N_k}{\partial t} \frac{d\vec{k}}{(2\pi)^3} = \frac{1}{2}\frac{\partial}{\partial t}(\vec{u}n_\gamma) - \int N_k \frac{\partial \vec{v}_k}{\partial t} \frac{d\vec{k}}{(2\pi)^3}. \qquad (4.100)$$

The contribution of the second term can be written as

$$\int \vec{v}_k \vec{v}_k \cdot \nabla N_k \frac{d\vec{k}}{(2\pi)^3} = \frac{1}{2}\nabla \cdot \left(n_\gamma \langle \vec{v}_k \vec{v}_k \rangle\right) - \int N_k \nabla \cdot (\vec{v}_k \vec{v}_k) \frac{d\vec{k}}{(2\pi)^3} \qquad (4.101)$$

where we have used the following quantity:

$$\langle \vec{v}_k \vec{v}_k \rangle = \frac{2}{n_\gamma} \int N_k \vec{v}_k \vec{v}_k \frac{d\vec{k}}{(2\pi)^3}. \qquad (4.102)$$

Here, we can introduce the concept of photon pressure, P_γ, such that

$$\langle \vec{v}_k \vec{v}_k \rangle = \vec{u}\vec{u} + \frac{P_\gamma}{n_\gamma}\mathbf{I} \qquad (4.103)$$

where $\mathbf{I}$ is the 3×3 identity matrix.

Using equation (4.102) and adding the diagonal terms of this tensorial relation, it can easily be shown that the photon pressure is determined by

$$P_\gamma = \frac{2}{3}\int N_k(v_k - u)^2 \frac{d\vec{k}}{(2\pi)^3}. \qquad (4.104)$$

We can also define an effective temperature for the photon gas, assuming that it can be considered an ideal gas, such that

$$P_\gamma = n_\gamma T_{\text{eff}}. \qquad (4.105)$$

Finally, the contribution from the third term of the photon kinetic equation to the momentum conservation law is determined by

$$\vec{v}_k \left(\vec{f} \cdot \frac{\partial N_k}{\partial \vec{k}}\right) \frac{d\vec{k}}{(2\pi)^3}. \qquad (4.106)$$

Adding the contributions from the three terms and using the continuity equation, we can finally establish the momentum conservation law of the photon gas, in the form

$$\frac{\partial \vec{u}}{\partial t} + \vec{u} \cdot \nabla \vec{u} = -\frac{\nabla P_\gamma}{n_\gamma} + \vec{F}_\gamma \qquad (4.107)$$

where the mean force acting on the photons is determined by

$$\vec{F}_\gamma = \frac{2}{n_\gamma}\int N_k \left[\frac{\partial \vec{v}_k}{\partial t} + \nabla \cdot (\vec{v}_k \vec{v}_k) + \frac{\partial}{\partial \vec{k}} \cdot (\vec{f}\vec{v}_k)\right] \frac{d\vec{k}}{(2\pi)^3}. \qquad (4.108)$$

As a first example, let us calculate this force for the case of a non-dispersive dielectric medium, with refractive index $n \equiv n(\vec{r}, t)$. In this case, we have

$$\omega = \frac{kc}{n}, \quad \vec{v}_k = \frac{c}{n}\frac{\vec{k}}{k}. \tag{4.109}$$

Using these in the above expression for $\vec{F}_\gamma$, we obtain

$$\vec{F}_\gamma = \frac{2}{n_\gamma}\frac{c^2}{n^2} \int N_k \left[-\frac{\partial n}{\partial t}\frac{\vec{k}}{k} + \frac{1}{n}\frac{\partial n}{\partial \vec{r}} \cdot \left(1 - 2\frac{\vec{k}\vec{k}}{k^2} \right) \right] \frac{d\vec{k}}{(2\pi)^3}. \tag{4.110}$$

We see that this mean force acting on the photon gas is due to the space and time variations of the refractive index. As a second example, let us consider an isotropic plasma, such that

$$\omega^2 = k^2 c^2 + \omega_p^2, \quad \vec{v}_k = \frac{\vec{k}c^2}{\omega}. \tag{4.111}$$

The mean force now becomes

$$\vec{F}_\gamma = -\frac{1}{n_\gamma} \int \frac{N_k}{\omega^2} \left(\nabla \omega_p^2 + \vec{v}_k \frac{\partial \omega_p^2}{\partial t} \right) \frac{d\vec{k}}{(2\pi)^3}. \tag{4.112}$$

If the distribution is even with respect to the quantity $(\vec{v}_k - \vec{u})$, this can also be written as

$$\vec{F}_\gamma = -\frac{U}{n_\gamma} \left(\nabla \omega_p^2 + \vec{u}\frac{\partial \omega_p^2}{\partial t} \right) \tag{4.113}$$

with

$$U = \int \frac{N_k}{\omega^2} \frac{d\vec{k}}{(2\pi)^3}. \tag{4.114}$$

In a similar way, we can also derive the energy conservation equation, by multiplying the kinetic equation by $\hbar\omega$ and integrating it over $\vec{k}$. The contribution of the first term is

$$\int \hbar\omega \frac{\partial N_k}{\partial t} \frac{d\vec{k}}{(2\pi)^3} = \frac{1}{2}\frac{\partial}{\partial t} W_\gamma - \hbar \int N_k \frac{\partial \omega}{\partial t} \frac{d\vec{k}}{(2\pi)^3} \tag{4.115}$$

where the mean photon energy is determined by

$$W_\gamma = \frac{2}{n_\gamma} \int \hbar\omega N_k \frac{d\vec{k}}{(2\pi)^3}. \tag{4.116}$$

The contribution of the second term is

$$\int \hbar\omega \vec{v}_k \cdot \nabla N_k \frac{d\vec{k}}{(2\pi)^3} = \frac{1}{2}\nabla \cdot (\vec{u}W_\gamma) - \hbar \int N_k \nabla \cdot (\omega \vec{v}_k) \frac{d\vec{k}}{(2\pi)^3}. \tag{4.117}$$

Finally, the contribution of the third term is

$$\int \hbar\omega \vec{f} \cdot \frac{\partial N_k}{\partial \vec{k}} \frac{d\vec{k}}{(2\pi)^3} = -\hbar \int N_k \frac{\partial}{\partial \vec{k}} \cdot (\omega \vec{f}) \frac{d\vec{k}}{(2\pi)^3} \qquad (4.118)$$

which exactly cancels the last term in equation (4.117).

Adding these three contributions, we obtain

$$\frac{\partial}{\partial t} W_\gamma + \nabla \cdot (\vec{u} W_\gamma) = 2\hbar \int N_k \frac{\partial \omega}{\partial t} \frac{d\vec{k}}{(2\pi)^3}. \qquad (4.119)$$

In the case of a non-dispersive dielectric medium, we have

$$\frac{\partial \omega}{\partial t} = -\frac{\omega}{n} \frac{\partial n}{\partial t} \qquad (4.120)$$

where n is the refractive index, and, in the case of a plasma,

$$\frac{\partial \omega}{\partial t} = \frac{1}{2\omega} \frac{\partial \omega_p^2}{\partial t} \qquad (4.121)$$

where ω_p is the electron plasma frequency.

We see that, in contrast with the momentum conservation equation (4.107, 4.108), where both the time and the space derivatives of the refractive index (or of the plasma frequency) contribute to the change in the mean value of the momentum of the photon gas, only the time derivatives can be a source of energy.

4.5 Self-phase modulation

An important property of the photon kinetic theory is that it can describe the nonlinear changes of a photon bunch or a short laser pulse due to its own spectrum. In other words, it can explain the following self-consistent process: the photon bunch changes the optical properties of the medium which, in turn, modify the spectral composition of the photon bunch. This was already shown in the example of the self-blueshift, given at the end of section 4.4.

The best example of such a process is however the well-known self-phase modulation of a laser pulse in an optical fibre or in other optical media. By this process, an initially nearly monochromatic laser pulse can be transformed into a broadband radiation pulse.

In more exact terms, if the initial laser pulse is a nearly transform limited short pulse, where the spectral width is close to its minimum value allowed by the uncertainty principle, the nonlinear process, usually called self-phase modulation, can transform it into nearly white light.

This astonishing effect has been known experimentally since the early 1970s and the current theory is based on the calculation of the nonlinear contributions to the total time phase dependence. However, the photon kinetic theory described in

this chapter suggests the possible use of a totally independent theoretical explanation, where the phase is completely ignored and where only the photon number
distribution is considered.

It will be shown here that this new approach is equally capable of describing
the spectral changes characterizing the so-called self-phase modulation processes,
which leads us to two important conclusions. The first one is that the phase is not
an essential ingredient of self-phase modulation. This is also valid for the induced
phase modulation, briefly discussed at the end of chapter 3. The second is that
self-phase modulation (as well as the other phase modulation processes, such as
induced and crossed phase modulation) is nothing but a particular aspect of the
phenomenology of photon acceleration described in this work.

Physically this means that, starting from a nearly monochromatic laser pulse,
the nonlinear properties of the optical media can accelerate and decelerate the
photons contained inside the pulse, spreading them over a considerable range
of the optical spectrum, thus leading to nearly white light. In order to clarify
these few very important physical statements, and to make them accessible to the
nonspecialist, let us first remind ourselves of the usual optical theory of self-phase
modulation and then compare it to the photon kinetic approach.

4.5.1   Optical theory

If we start from Maxwell's equations in a dielectric medium, in the absence of
charge and current distributions, we can easily derive the following equation of
propagation for the electric field $\vec{E}$ associated with the laser pulse:

$$\nabla^2 \vec{E} - \nabla(\nabla \cdot \vec{E}) - \frac{1}{c^2}\frac{\partial^2 \vec{E}}{\partial t^2} = \mu_0 \frac{\partial^2 \vec{P}}{\partial t^2}. \tag{4.122}$$

The vector ploarization appearing in this equation can be divided into two
distinct parts: $\vec{P} = \vec{P}_L + \vec{P}_{NL}$. The linear part is determined by

$$\vec{P}_L(t) = \epsilon_0 \int_0^\infty \chi^{(1)}(\tau)\vec{E}(t - \tau)\, d\tau. \tag{4.123}$$

Here we assume that the medium is isotropic, otherwise the linear (or first-
order) susceptibility $\chi^{(1)}$ would be replaced by a tensor. For the nonlinear part of
the polarization vector, we can neglect dispersion and simply write

$$\vec{P}_{NL}(t) = \epsilon_0 \chi^{(3)}|E(t)|^2 \vec{E}(t). \tag{4.124}$$

Assuming that the laser field is transverse ($\nabla \cdot \vec{E} = 0$), we can then write the
propagation equation as

$$\nabla^2 \vec{E} - \frac{1}{c^2}\frac{\partial^2}{\partial t^2}\left\{\vec{E} + \int_0^\infty \chi^{(1)}(\tau)\vec{E}(t - \tau)\, d\tau\right\} = \frac{1}{c^2}\frac{\partial^2}{\partial t^2}\chi^{(3)}|E|^2\vec{E}. \tag{4.125}$$

Let us assume a solution of the form

$$\vec{E}(\vec{r}, t) = \frac{1}{2}\left[\vec{a}\, E_0(z, t)\exp(ik_0z - i\omega_0 t) + \text{c.c.}\right].\qquad(4.126)$$

This represents a laser pulse propagating along the z-axis, with a unit polarization vector $\vec{a}$ and a slowly varying amplitude $E_0(z, t)$. This means that the scales of variation of this amplitude are such that

$$\left|\frac{1}{E_0}\frac{\partial E_0}{\partial t}\right| \ll \omega_0, \qquad \left|\frac{1}{E_0}\frac{\partial E_0}{\partial z}\right| \ll k_0.\qquad(4.127)$$

We can then write

$$\frac{\partial^2 \vec{E}}{\partial t^2} \simeq -\omega_0^2\vec{E} + \frac{1}{2}\left[-2i\omega_0\vec{a}\exp(ik_0z - i\omega_0 t)\frac{\partial E_0}{\partial t} + \text{c.c.}\right]\qquad(4.128)$$

$$\nabla^2\vec{E} \simeq -k_0^2 + \frac{1}{2}\left[2ik_0\vec{a}\exp(ik_0z - i\omega_0 t)\frac{\partial E_0}{\partial z} + \text{c.c.}\right].$$

We will also assume weak linear dispersion, which means that the function $\chi^{(1)}(\tau)$ is sharply peaked around $\tau = 0$. The absence of dispersion would correspond to identifying this function with a Dirac delta function $\delta(\tau)$.

In the weak dispersion medium, we can expand the electric field $\vec{E}(t - \tau)$ appearing in equation (4.123) as

$$\vec{E}(t - \tau) \simeq \vec{E}(t) - \tau\frac{\partial}{\partial t}\vec{E}(t) + \cdots.\qquad(4.129)$$

Using this expansion in equation (4.123), and noting that

$$\chi(\omega) = \int_0^\infty \chi^{(1)}(t)e^{i\omega t}\,dt\qquad(4.130)$$

$$\frac{\partial}{\partial\omega}\chi(\omega) = i\int_0^\infty t\chi^{(1)}(t)e^{i\omega t}\,dt$$

we obtain

$$\vec{P}_{NL}(t) = \frac{\epsilon_0}{2}\left\{\vec{a}\exp(ik_0z - i\omega_0 t)\left[\chi(\omega_0) + i\frac{\partial\chi(\omega_0)}{\partial\omega}\frac{\partial}{\partial t}\right]E_0(z, t) + \text{c.c.}\right\}.\qquad(4.131)$$

We can now use equations (4.129, 4.131) in the propagation equation (4.125). Noting that the frequency ω_0 and the wavenumber k_0 are related by the linear dispersion relation $k^2c^2 = \omega^2\epsilon(\omega)$, where $\epsilon(\omega) = 1 + \chi(\omega)$ is the dielectric function of the medium, we obtain an equation describing the slow space–time evolution of the envelope field amplitude:

$$\left(\frac{\partial}{\partial z} + \frac{1}{v_0}\frac{\partial}{\partial t}\right)E_0 = i\omega_0\alpha|E_0|^2 E_0.\qquad(4.132)$$

Here v_0 is the group velocity, determined by

$$\frac{1}{v_0} = \frac{k_0}{\omega_0} + \frac{\omega_0^2}{2c^2}\frac{\partial\epsilon(\omega_0)}{\partial\omega} \tag{4.133}$$

and α is the nonlinear coefficient, resulting from the existence of a third-order susceptibility

$$\alpha = \frac{\omega_0}{k_0}\frac{1}{c^2}\chi^{(3)}. \tag{4.134}$$

The solution of the envelope equation (4.132) can be written in the form

$$E_0(z,t) = A(\eta)\,e^{i\phi(\eta,t)} \tag{4.135}$$

where we have $\eta = z - v_0 t$ and the nonlinear phase

$$\phi(\eta,t) = \phi_0 + \omega_0\alpha|A(\eta)|^2 t. \tag{4.136}$$

Such a solution describes a laser pulse propagating with an invariant pulse shape, but with a phase which depends on time (or on the travelled distance), as well as on the form of that shape. The result is that the pulse frequency will not remain constant along the propagation and will depend on the actual position inside the pulse envelope:

$$\omega = \omega_0 - \frac{\partial\phi}{\partial t} = \omega_0\left[1 + \alpha\frac{t}{v_0}\frac{\partial}{\partial\eta}|A(\eta)|^2\right]. \tag{4.137}$$

As shown before, this space–time dependence of the frequency (or of the wavevector) is called the pulse chirp. In order to have a more precise idea of the chirp, and of the amount of frequency shift $\Delta\omega = \omega - \omega_0$ introduced by the nonlinear susceptibility of the medium, we can take the simple and illustrative example of a Gaussian laser pulse, such that

$$|A(\eta)|^2 = A_0^2\,e^{-\eta^2/\sigma^2}. \tag{4.138}$$

From equation (4.137), we get

$$\Delta\omega = -\alpha t\frac{\omega_0}{v_0}\frac{2\eta}{\sigma^2}|A(\eta)|^2. \tag{4.139}$$

We see that the frequency shift grows linearly with time, and that it is negative at the pulse front, for $\eta > 0$, and positive at the rear, for $\eta < 0$.

4.5.2 Kinetic theory

An alternative description of the same effect is now given in terms of photon acceleration [103]. Instead of using the envelope field equation, we use, as our starting point, the equation describing the space–time evolution of the photon number distribution $N_k(\vec{r}, t)$, which can be written as

$$\frac{d}{dt} N_k(\vec{r}, t) = 0 \tag{4.140}$$

where the total time derivative is

$$\frac{d}{dt} = \frac{\partial}{\partial t} + \frac{d\vec{r}}{dt} \cdot \frac{\partial}{\partial \vec{r}} + \frac{d\vec{k}}{dt} \cdot \frac{\partial}{\partial \vec{k}}. \tag{4.141}$$

In order to determine this total time derivative operator we have to make use of the ray equations

$$\frac{d\vec{r}}{dt} = \frac{\partial \omega}{\partial \vec{k}}, \quad \frac{d\vec{k}}{dt} = -\frac{\partial \omega}{\partial \vec{k}} \tag{4.142}$$

where, in the expression of the frequency $\omega \equiv \omega_k(\vec{r}, t)$, we retain the nonlinear corrections to the refractive index

$$\omega = \frac{kc}{n} = \frac{kc}{n_0 + n_2 I(\vec{r}, t)}. \tag{4.143}$$

Here, n_0 is the linear refractive index, n_2 is the nonlinear one and $I(\vec{r}, t)$ is the laser pulse intensity, where

$$I(\vec{r}, t) = \hbar \int \omega_k N_k(\vec{r}, t) \frac{d\vec{k}}{(2\pi)^3}. \tag{4.144}$$

This is a slowly varying function as compared with the frequency timescale:

$$\omega \gg \left| \frac{\partial}{\partial t} \ln I(\vec{r}, t) \right|. \tag{4.145}$$

Using equation (4.143) in (4.142), we get

$$\frac{d\vec{r}}{dt} = \frac{c}{n_0 + n_2 I(\vec{r}, t)} \tag{4.146}$$

$$\frac{d\vec{k}}{dt} = \frac{kc}{[n_0 + n_2 I(\vec{r}, t)]^2} n_2 \frac{\partial}{\partial \vec{r}} I(\vec{r}, t). \tag{4.147}$$

We know that the formal solution of the photon kinetic equation (4.140) can be written as

$$N(\vec{r}, \vec{k}, t) = N(\vec{k}_0(\vec{r}, \vec{k}, t), \vec{r}_0(\vec{r}, \vec{k}, t), t_0) \tag{4.148}$$

where $(\vec{k}_0, \vec{r}_0, t_0)$ are the initial conditions.

This means that, by solving equations (4.146, 4.147) we are able to determine the evolution of $N_k(\vec{r}, t)$. In order to illustrate the procedure, let us solve these equations in a one-dimensional situation, where propagation is taken along the z-axis.

We will also assume that the nonlinear corrections to the refractive index are small, $n_0 \gg n_2 I(\vec{r}, t)$, which is usually the case, even for very intense laser pulses. We can then reduce equations (4.146, 4.147) to

$$\frac{dz}{dt} = \frac{c}{n_0}\left[1 - \frac{n_2}{n_0}I(z, t)\right] \tag{4.149}$$

$$\frac{dk}{dt} = \frac{kc}{n_0}\frac{n_2}{n_0}\frac{\partial}{\partial z}I(z, t). \tag{4.150}$$

Let us also assume that the pulse propagates without significant profile deformation, with a group velocity c/n_0. This means that

$$I(z, t) \simeq I\left(z - \frac{c}{n_0}t\right). \tag{4.151}$$

This suggests the use of a canonical transformation from (z, k) to a new pair of variables (η, p), such that

$$\eta = z - \frac{c}{n_0}t, \quad p = k. \tag{4.152}$$

The resulting Hamiltonian function is

$$\Omega(\eta, p, t) = \omega(\eta, p, t) - p\frac{c}{n_0}. \tag{4.153}$$

Using the approximate expression

$$\omega \simeq \frac{kc}{n_0}\left[1 - \frac{n_2}{n_0}I(z, t)\right] = \frac{pc}{n_0}\left[1 - \frac{n_2}{n_0}I(\eta)\right] \tag{4.154}$$

we get

$$\Omega(\eta, p) = -pwI(\eta) \tag{4.155}$$

where we have used

$$w = \frac{n_2}{n_0}\frac{c}{n_0}. \tag{4.156}$$

The ray equations can now be written in the form

$$\frac{d\eta}{dt} = \frac{\partial\Omega}{\partial p} = -wI(\eta), \quad \frac{dp}{dt} = -\frac{\partial\Omega}{\partial \eta} = pw\frac{\partial I(\eta)}{\partial \eta}. \tag{4.157}$$

The integration of these photon equations of motion is straighforward, and the result can be written as

$$\int_{\eta_0}^{\eta}\frac{d\eta'}{I(\eta')} = -w(t - t_0) \tag{4.158}$$

and

$$p_0 = p \exp\left[-w \int_{t_0}^{t} \frac{\partial I(\eta)}{\partial \eta} \, \mathrm{d}t\right]. \tag{4.159}$$

The first of these expressions shows that the initial value of the position coordinate η is independent of the momentum variable p:

$$\eta_0 \equiv \eta_0(\eta, t). \tag{4.160}$$

With regard to the second expression, it corresponds to an implicit integration, but it will be very useful for the calculation of the spectral changes of the laser pulse along its propagation. Let us show this by considering the chirp of the pulse $\langle\omega\rangle$.

By definition, the chirp is the averaged frequency of the pulse, at a given position η and at a given instant t:

$$\langle\omega\rangle_{\eta,t} = \int N_\mathrm{p}(\eta, t)\omega(\eta, p, t) \, \mathrm{d}p. \tag{4.161}$$

Using equations (4.148, 4.154), we get

$$\langle\omega\rangle_{\eta,t} = \int \frac{pc}{n_0}\left[1 - \frac{n_2}{n_0}I(\eta)\right] N(\eta_0, p_0, t_0) \, \mathrm{d}p. \tag{4.162}$$

Using equation (4.159) we can write $p \, \mathrm{d}p$ in terms of $p_0 \, \mathrm{d}p_0$, which leads to

$$\langle\omega\rangle_{\eta,t} = \int \frac{p_0 c}{n_0}\left[1 - \frac{n_2}{n_0}I(\eta)\right] N(\eta_0, p_0, t_0) \exp\left[2w \int \frac{\partial I(\eta)}{\partial \eta} \, \mathrm{d}t\right]. \tag{4.163}$$

But for a non-dispersive pulse, we have $I(\eta) = I(\eta_0)$, as already stated in equation (4.151). This means that we can write the pulse chirp in the form

$$\langle\omega\rangle_{\eta,t} = \langle\omega\rangle_0 \exp\left[2w \frac{\partial I(\eta)}{\partial \eta}(t - t_0)\right] \tag{4.164}$$

where $\langle\omega\rangle_0 \equiv \langle\omega\rangle_{\eta_0, t_0}$ is the initial pulse chirp.

This is a simple but important result, which will allow us to rediscover the main features of the so-called self-phase modulation, this time without considering the field phase. In particular, we can determine the condition for an extremum

$$\frac{\partial}{\partial \eta}\langle\omega\rangle_{\eta,t} = 0. \tag{4.165}$$

This will correspond to the stationary points $\langle\omega\rangle_{\max}$ of the chirp curve. From equation (4.164) it is obvious that

$$\frac{\partial}{\partial \eta}\langle\omega\rangle_{\eta,t} = 2w(t - t_0)\langle\omega\rangle_{\eta,t}\left[\frac{\partial^2}{\partial \eta^2}I(\eta)\right] = 0. \tag{4.166}$$

From this we conclude that the point $\eta = \eta_{max}$, for which the maximum value of the chirp $\langle \omega \rangle_{max}$ is observed, is determined by the condition

$$\frac{\partial^2}{\partial \eta^2} I(\eta) = 0. \tag{4.167}$$

Let us illustrate these results by assuming a Gaussian pulse

$$I(\eta) = I_0 e^{-\eta^2/\sigma^2}. \tag{4.168}$$

For this particular pulse shape, we have

$$\frac{\partial^2}{\partial \eta^2} I(\eta) = \left(\frac{2\eta^2}{\sigma^2} - 1 \right) \frac{2}{\sigma^2} I(\eta). \tag{4.169}$$

Comparing this with condition (4.167), we conclude that the chirp maxima are located at the two points determined by $2\eta^2 = \sigma^2$, or equivalently, by

$$\eta_{max} = \pm \frac{1}{\sqrt{2}} \sigma. \tag{4.170}$$

From equation (4.168) we then get

$$\left. \frac{\partial}{\partial \eta} I(\eta) \right|_{\eta_{max}} = \mp \frac{\sqrt{2}}{\sigma} I_0 e^{-1/2}. \tag{4.171}$$

Using equation (4.164) we finally obtain the condition for the maximum chirp:

$$\langle \omega \rangle_{max} = \langle \omega \rangle_0 \exp \left[\mp \frac{2\sqrt{2}}{\sigma} w I_0 e^{-1/2} (t - t_0) \right]. \tag{4.172}$$

It should be noticed that this concept of maximum chirp can also be physically identified with the condition for the maximum frequency shift with respect to the initial pulse frequency. Introducing the frequency shift as $\Delta \omega = \langle \omega \rangle_{max} - \langle \omega \rangle_0$, we get, for small arguments of the above exponential,

$$\Delta \omega \simeq \pm \frac{2\sqrt{2}}{\sigma} \frac{n_2}{n_0} \frac{c}{n_0} I_0 e^{-1/2} \Delta t \tag{4.173}$$

where we have used $\Delta t = t - t_0$.

Notice that such an expansion is valid for small time intervals, or for small nonlinearities. This result for the pulse frequency shift is in agreement (not only qualitatively, but also quantitatively) with that obtained above for the self-phase modulation of the laser pulse in a nonlinear medium: the spectral width grows linearly with time and the maxima down-shifts and the up-shifts are equal to each other.

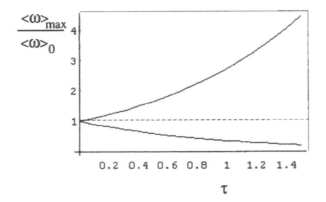

Figure 4.3. Time evolution of the maximum frequency shift associated with self-phase modulation, as determined by the photon kinetic theory.

We should however notice that, for long time intervals $\Delta t \rightarrow \infty$, the above expansion in not valid and the Stokes and anti-Stokes sidebands become quite asymmetric. This is clearly seen in the more exact equation (4.172) where, for long time intervals, one of the branches of the exponential tends to zero, while the other tends to infinity. Such a spectral feature is well confirmed by experiments and is illustrated in figure 4.3.

This asymmetry is a natural consequence of the present theoretical formulation, even for an initially symmetric laser pulse. In contrast, it cannot be derived from the optical theory of self-phase modulation discussed above.

Another important feature of this effect, which is the pulse steepening, can also be explained by our kinetic model. In order to study the changes in the pulse shape, we can no longer use the non-dispersive solution stated in equation (4.151), which was one of the basic assumptions of our analytical calculation of the pulse chirp. Instead, we have to solve numerically the photon kinetic equation. The result is illustrated in figure 4.4, where the pulse steepening, resulting from the group (or photon) velocity dispersion, is clearly shown.

We conclude this section by stating that the effect usually called self-phase modulation can be accurately described with the photon kinetic equation, where the wavefield phase is completely ignored. The three main experimental features of this effect are well described by this kinetic model: the pulse chirp, the spectral asymmetry between the Stokes and the anti-Stokes sidebands, and the steepening of the pulse shape. In many respects, the analytical solutions of the kinetic theory are more accurate than the analytical solutions of the optical theory, in particular for the prediction of spectral asymmetry.

We are then led to two different kinds of conclusion. The first one is that the so-called self-phase modulation can be seen as a particular example of the more general concept of photon acceleration in a non-stationary medium. In this

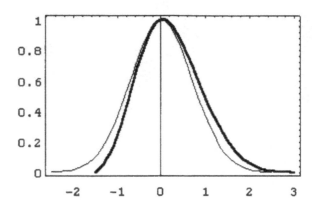

Figure 4.4. Illustration of pulse steepening associated with self-phase modulation, as described by the kinetic theory. A Gaussian pulse is also shown, for comparison.

particular case, the non-stationarity of the medium is due to the laser pulse itself (or to the photon number distribution associated with the pulse) which locally changes the refractive index, due to the nonlinear response of the medium.

The second, and more surprising, conclusion is that the wavefield phase is not the essential ingredient of this effect: two laser pulses, with different phase contents, would lead to nearly the same spectral broadening in the same medium. This means that the explanation of such a frequency shift or spectral change is conceptually more accurate in terms of photon kinetics than in terms of phase modulation.

Chapter 5

Photon equivalent charge

Until now, we have only considered the linear properties of photons in a medium, which essentially derive from their effective mass. In some sense these are the most relevant ones.

In this chapter we turn our attention to the influence of the nonlinear properties of an optical medium, and we show that the photons possess an equivalent electric charge. In a plasma, this equivalent charge can be directly associated with the ponderomotive force (or the radiation pressure) of the photon gas acting on the plasma electrons. We will later see that in a non-ionized medium, for instance an optical fibre, this electric charge is replaced by an electric dipole.

This new, and somewhat unexpected property of a photon is due to the space–time varying polarization effects that it induces in the medium. We can then say that a photon in a vacuum is a 'bare' particle with no electric charge, but a photon with the same frequency (or energy) travelling in a plasma has to be considered a 'dressed' particle with an electric charge which is not quantized and depends on the plasma density [69, 104, 112]. Furthermore, the existence of this charge allows us to formulate a new series of problems, related to possible radiation mechanisms due to accelerated 'dressed' photons, or to their motion across the boundary between two different media.

Finally, a new property associated with this electric charge (or in other words, with the radiation pressure effects) is the possibility of resonant Cherenkov interactions between photons and electron plasma waves, leading to the concept of photon Landau damping [14]. Plasma instabilities, induced by particular photon spectral distributions (such as photon beams), can then be envisaged.

5.1 Derivation of the equivalent charge

One possible way to derive the photon equivalent charge is to consider an electromagnetic wavepacket propagating in a plasma. Once again, we will assume for simplicity that the plasma is unmagnetized. The electric field $\vec{E} \equiv \vec{E}(\vec{r}, t)$

associated with this wavepacket is described by the wave equation

$$\nabla^2 \vec{E} - \nabla(\nabla \cdot \vec{E}) - \frac{1}{c^2}\frac{\partial^2 \vec{E}}{\partial t^2} = \mu_0 \frac{\partial \vec{J}}{\partial t}. \tag{5.1}$$

For high-frequency waves, we can neglect the ion motion, and the current in this equation is reduced to the electron current

$$\vec{J} = -en\vec{v}. \tag{5.2}$$

Phase velocities of transverse electromagnetic waves in an unmagnetized plasma are always larger than the speed of light c, which means that kinetic effects are usually negligible and that we can use the electron fluid equations in order to determine the electron density n and the electron mean velocity.

In their simple non-relativistic version (which is valid only for moderate wave intensities) the electron fluid equations are

$$\frac{\partial n}{\partial t} + \nabla \cdot n\vec{v} = 0 \tag{5.3}$$

$$\frac{\partial \vec{v}}{\partial t} + \vec{v} \cdot \nabla \vec{v} = -\frac{e}{m}\left(\vec{E} + \vec{v} \times \vec{B}\right). \tag{5.4}$$

In the first of these equations we recognize the continuity equation of the electron fluid, and in the second one, the momentum conservation equation. The magnetic field $\vec{B}$ appearing in the Lorentz force term is the magnetic field of the electromagnetic wave itself and can be expressed in terms of the electric field $\vec{E}$ by using the Faraday equation

$$\nabla \times \vec{E} = -\frac{\partial \vec{B}}{\partial t}. \tag{5.5}$$

Equations (5.4) are nonlinear and their solution is in general quite complicated. However we can, as a first approximation, linearize them with respect to the wave electric field. Let us call $\vec{E}_1$ this linear solution for the electric field.

If we restrict our discussion to purely transverse waves, we have $\nabla \cdot \vec{E}_1 = 0$ and the wave equation (5.1) reduces to

$$\nabla^2 \vec{E}_1 - \frac{1}{c^2}\frac{\partial^2 \vec{E}_1}{\partial t^2} = \mu_0 \frac{\partial \vec{J}_1}{\partial t} \tag{5.6}$$

where the linear current is

$$\vec{J}_1 = -en_0\vec{v}_1. \tag{5.7}$$

Here n_0 is the equilibrium (or unperturbed) electron plasma density. The linear response of the plasma is determined by the linearized version of equations (5.3, 5.4)

$$\frac{\partial n_1}{\partial t} + n_0\nabla \cdot \vec{v} = 0 \tag{5.8}$$

$$\frac{\partial \vec{v}_1}{\partial t} = -\frac{e}{m}\vec{E}_1. \tag{5.9}$$

The solution of this equation can take the general form

$$\vec{E}_1(\vec{r}, t) = \frac{1}{2}\vec{E}_0(\vec{r}, t)\exp(i\vec{k}_0 \cdot \vec{r} - i\omega_0 t) + \text{c.c.} \tag{5.10}$$

where ω_0 and $\vec{k}_0$ are the mean frequency and wavevector of the wavepacket and $\vec{E}_0(\vec{r}, t)$ is a slowly varying amplitude. We should notice that such a general expression could also be a solution for nonlinear wavepackets.

According to equations (5.8, 5.9), the electron density and velocity perturbations associated with this solution have similar expressions, with amplitudes determined by

$$n_1 = n_0\frac{\vec{k}_0}{\omega_0} \cdot \vec{v}_1, \quad \vec{v}_1 = \frac{e}{i\omega_0 m}\vec{E}_0. \tag{5.11}$$

We notice that, for purely tranverse wave solutions, such that $\vec{k}_0 \cdot \vec{E}_0 = 0$, the electron density perturbations are equal to zero: $n_1 = 0$. Finally, the amplitude of the wave magnetic field is, according to equation (5.5),

$$\vec{B}_1 = \frac{\vec{k}_0}{\omega_0} \times \vec{E}_0. \tag{5.12}$$

Once this simple linear solution is understood, we can return to the wave equation (5.1) and try to obtain a more adequate solution which takes the nonlinear terms into account. This can be done by assuming that the total electric field in this equation is divided in two parts: the main field (which corresponds to the linear solution) and a small correction due to radiation or polarization effects, i.e.

$$\vec{E} = \vec{E}_1 + \vec{E}_2 \tag{5.13}$$

with

$$|\vec{E}_1| \gg |\vec{E}_2|. \tag{5.14}$$

To the first order in the small nonlinear field $\vec{E}_2$, we can obtain from equation (5.1)

$$\nabla^2\vec{E}_2 - \nabla(\nabla \cdot \vec{E}_2) - \frac{1}{c^2}\frac{\partial^2}{\partial t^2}\vec{E}_2 = \mu_0\frac{\partial}{\partial t}\vec{J}_2 \tag{5.15}$$

where the nonlinear current is determined by

$$\vec{J}_2 = -en_0\vec{v}_2. \tag{5.16}$$

To the same order of approximation, and assuming that equations (5.8, 5.9), stay valid, we can also get from equations (5.3, 5.4),

$$\frac{\partial n_2}{\partial t} + n_0\nabla \cdot \vec{v}_2 = 0 \tag{5.17}$$

$$\frac{\partial \vec{v}_2}{\partial t} + \vec{v}_1 \cdot \nabla\vec{v}_1 = -\frac{e}{m}(\vec{E}_2 + \vec{v}_1 \times \vec{B}_1). \tag{5.18}$$

Now we should make use of the vector identity

$$\nabla(\vec{a} \cdot \vec{b}) = (\vec{a} \cdot \nabla)\vec{b} + (\vec{b} \cdot \nabla)\vec{a} + \vec{a} \times (\nabla \times \vec{b}) + \vec{b} \times (\nabla \times \vec{a}). \qquad (5.19)$$

If we identify the vector $\vec{a}$ with the first-order velocity perturbation $\vec{v}_1$ and the vector $\vec{b}$ with its complex conjugate, we can easily realize that equation (5.18) can be rewritten as

$$\frac{\partial \vec{v}_2}{\partial t} + \frac{1}{2}\nabla|v_1|^2 = -\frac{e}{m}\vec{E}_2. \qquad (5.20)$$

In doing so we are neglecting the high-frequency part of the nonlinear perturbation $\vec{v}_2$, which would not be valid if we wanted to study harmonic generation. After elimination of $\vec{v}_2$ from equations (5.17, 5.20) we get the following evolution equation for the low-frequency nonlinear perturbation of the electron plasma density:

$$\frac{\partial^2 n_2}{\partial t^2} - \frac{en_0}{m}\nabla \cdot \vec{E}_2 = \frac{n_0}{2}\nabla^2|v_1|^2. \qquad (5.21)$$

Using equation (5.11) this can also be written as the equation of a forced linear oscillator

$$\frac{\partial^2 n_2}{\partial t^2} + \omega_p^2 n_2 = \frac{1}{2}\frac{e^2 n_0}{m^2 \omega_0^2}\nabla^2|E_0(\vec{r}, t)|^2. \qquad (5.22)$$

The eigenfrequency of this linear oscillator is nothing but the unperturbed electron plasma frequency $\omega_p = (e^2 n_0/\epsilon_0 m)^{1/2}$. In this equation $|E_0(\vec{r}, t)|^2$ describes the slowly varying envelope of the electromagnetic wavepacket.

If this wavepacket propagates along the x-axis, with a group velocity $\vec{v}_g = v_g\vec{e}_x$, with negligible shape deformation, we can write

$$\vec{E}_0(\vec{r}, t) \equiv \vec{E}_0(\xi, \vec{r}_\perp) \qquad (5.23)$$

with

$$\xi = x - v_g t. \qquad (5.24)$$

We can now perform a coordinate transformation from (x, t) to (ξ, τ), with $\tau = t$. This means that

$$\frac{\partial}{\partial x} = \frac{\partial}{\partial \xi}, \quad \frac{\partial}{\partial t} = \frac{\partial}{\partial \tau} - v_g\frac{\partial}{\partial \xi}. \qquad (5.25)$$

Using this in the equation for the forced oscillator (5.22), we obtain

$$\left(\frac{\partial^2}{\partial \tau^2} + \omega_p^2 + v_g^2\frac{\partial^2}{\partial \xi^2} - 2v_g\frac{\partial^2}{\partial \tau \partial \xi}\right) n_2$$

$$= \frac{\epsilon_0}{2m}\left(\frac{\omega_p}{\omega_0}\right)^2 \left(\frac{\partial^2}{\partial \xi^2} + \nabla_\perp^2\right) |E_0(\xi, \vec{r}_\perp)|^2. \qquad (5.26)$$

Before solving this equation we can simplify it in two ways. First we notice that the gradients in the perpendicular direction can be considered negligible in several relevant physical situations. In particular, for a 'pancake'-like wavepacket (a very short laser pulse), or for wavepackets not very much different from plane waves, we can always make $\nabla_\perp^2 \ll \partial^2/\partial\xi^2$.

Second, if we assume that, in the frame of the wavepacket the pulse shape and the electron perturbations induced by it are not significantly changing, we can also use the so-called quasi-static approximation: $\partial/\partial\tau \simeq 0$. With these two drastic but physically acceptable approximations the above equation is reduced to a much simpler form:

$$\left(\frac{\partial^2}{\partial\xi^2} + k_p^2\right) n_2 = \frac{\epsilon_0}{2m} \frac{k_p^2}{\omega_0^2} \frac{\partial^2}{\partial\xi^2} |E_0(\xi)|^2 \tag{5.27}$$

with $k_p = \omega_p/v_g$.

It is useful, at this point, to introduce a dimensionless space variable $\eta = k_p\xi$ and to replace the square module of the electric field by the number of photons

$$N(\xi) = \frac{\epsilon}{4\hbar\omega_0} |E_0(\xi)|^2. \tag{5.28}$$

Equation (5.27) now becomes

$$\left(\frac{\partial^2}{\partial\eta^2} + 1\right) n_2 = f(\eta) \tag{5.29}$$

where the force term is

$$f(\eta) = \frac{2\hbar k_p^2}{m\omega_0} \frac{\partial^2}{\partial\eta^2} N(\eta). \tag{5.30}$$

An appropriate solution of equation (5.29) is

$$n_2(\eta) = \int_\infty^\eta f(\eta') \sin(\eta - \eta') \, d\eta'. \tag{5.31}$$

Integrating by parts, we have

$$n_2(\eta) = \left[g(\eta') \sin(\eta - \eta')\right]_\infty^\eta + \int_\infty^\eta g(\eta') \cos(\eta - \eta') \, d\eta'. \tag{5.32}$$

Here we have used $f(\eta) = dg/d\eta$. Noting that $g(\eta') = 0$ for $\eta' \to \infty$, and that $\sin(\eta - \eta') = 0$ for $\eta' = \eta$, we obtain

$$n_2(\eta) = \int_\infty^\eta g(\eta') \cos(\eta - \eta') \, d\eta' \tag{5.33}$$

or, more explicitly,

$$n_2(\eta) = \frac{2\hbar k_p^2}{m\omega_0} \int_\infty^\eta \cos(\eta - \eta') \frac{\partial}{\partial\eta'} N(\eta') \, d\eta'. \tag{5.34}$$

This equation is appropriate for a second integration by parts. But for our purpose this will not be necessary because we notice that, in the limit of very short pulses, such that the pulse duration Δt is shorter than the period of the electron plasma oscillations $\Delta t \ll \omega_p^{-1}$, we have $\Delta \eta = (\eta - \eta') \ll 1$. This means that we can take $\cos(\eta - \eta') \simeq 1$, and obtain

$$n_2(\eta) = \frac{2\hbar k_p^2}{m\omega_0} N(\eta). \tag{5.35}$$

We can now define a photon equivalent charge Q_{ph}, such that

$$-en_2(\eta) = Q_{ph} N(\eta). \tag{5.36}$$

This means that

$$Q_{ph} = -2\hbar \frac{ek_p^2}{m\omega_0}. \tag{5.37}$$

The current associated with the $N(\eta)$ particles of charge Q_{ph} moving with velocity $\vec{v}_g = v_g \vec{e}_x$ is obviously given by

$$\vec{J}_{2NL} = Q_{ph} N(\eta) v_g \vec{e}_x. \tag{5.38}$$

The same expression for the second-order nonlinear current can also be derived if we start from the time derivative of the total second-order current $\vec{J}_2 = \vec{J}_{2L} + \vec{J}_{2NL}$:

$$\frac{\partial \vec{J}_2}{\partial t} = -en_0 \frac{\partial \vec{v}_2}{\partial t} = \frac{e^2 n_0}{m} \vec{E}_2 + \frac{\epsilon_0}{2m} \frac{\omega_p^2}{\omega_0^2} \nabla |E_0|^2. \tag{5.39}$$

Writing this equation in terms of the variables (ξ, τ) and making the above two assumptions of quasi-static $(\partial/\partial \tau = 0)$ and of a 'pancake' pulse $(\nabla_\perp \ll \partial/\partial \xi)$ approximations, we get

$$-v_g \frac{\partial \vec{J}_2}{\partial \xi} = \frac{e^2 n_0}{m} \vec{E}_2 - v_g^2 \frac{\partial}{\partial \xi} Q_{ph} N(\xi) \vec{e}_x. \tag{5.40}$$

We see that the second term leads to an expression for $\vec{J}_{2NL}$ which coincides with equation (5.38). This justifies the use of the concept of photon equivalent charge introduced above.

Our derivation of Q_{ph} was simplified by considering the integration of equation (5.34) in the limit of a very short pulse or wavepacket. But it can also be obtained if we consider a pulse with an arbitrary duration. In this case, the electron second-order perturbation will consist of two terms: the above term (5.35) which is directly related to the photon effective charge Q_{ph}, and a second term which represents the wakefield or plasma oscillation left behind the pulse [69].

To illustrate the existence of these two components, let us return to equation (5.34) and integrate it by parts. We are led to

$$n_e(\eta) = \frac{2\hbar k_p^2}{m\omega_0}\left[N(\eta) - \int_\infty^\eta N(\eta')\sin(\eta - \eta')\,d\eta'\right] \tag{5.41}$$

or, using equation (5.37),

$$n_2(\eta) = -\frac{1}{e}Q_{ph}N(\eta) + \frac{1}{e}Q_{ph}\int_\infty^\eta N(\eta')\sin(\eta - \eta')\,d\eta'. \tag{5.42}$$

The first term was already given by equation (5.36) and represents the equivalent charge of the total photon distribution. The second term represents the wakefield left behind it. It is nothing but an electron plasma oscillation moving with a phase velocity equal to the photon velocity v_g, and which follows the pulse as a wake. This justifies the statement made in chapter 3 about the possibility of generating relativistic electron plasma waves by means of a short laser pulse propagating in a plasma.

Another interesting aspect of the above calculations is the negative value of the photon effective charge. The sign of Q_{ph} can be physically well understood if we notice that the ponderomotive force appearing on the right-hand side of equation (5.21) tends to push the plasma electrons out of the region occupied by the electromagnetic wavepacket. This means that the photons tend to repel the electrons. Their equivalent charge is therefore negative.

5.2 Photon ondulator

We have just seen how the nonlinear current associated with an electromagnetic wavepacket propagating in a plasma can be derived. Our result (5.37, 5.38) shows that such a current is due to a flow of a number $N(\xi)$ of photons with charge Q_{ph} moving with velocity $\vec{v}_g$. Replacing this in the wave equation (5.15) we obtain

$$\nabla^2 \vec{E}_2 - \nabla(\nabla \cdot \vec{E}_2) - \frac{1}{c^2}\frac{\partial^2}{\partial t^2}\vec{E}_2 = \frac{\omega_p^2}{c^2} + \mu_0\frac{\partial}{\partial t}Q_{ph}N(\xi)\vec{v}_g. \tag{5.43}$$

This equation determines the secondary field radiated by the localized wavepacket (or the primary photon bunch) moving across the medium. An example of such a radiation effect is discussed here. Let us restrict our discussion to purely transverse fields, such that $\nabla \cdot \vec{E}_2 = 0$.

Using a time Fourier transformation

$$\vec{E}_2(\vec{r}, t) = \int_{-\infty}^{\infty} \vec{E}_\omega(\vec{r})\,e^{-i\omega t}\,\frac{d\omega}{2\pi} \tag{5.44}$$

we obtain

$$\nabla^2\vec{E}_\omega + \frac{\omega^2}{c^2}\epsilon(\omega)\vec{E}_\omega = -i\omega\mu_0\int_{-\infty}^{\infty}\vec{v}_g Q_{ph}N(\xi)\,e^{i\omega t}\,dt. \tag{5.45}$$

Here, we have used the plasma dielectric constant $\epsilon(\omega) = 1 - \omega_p^2/\omega^2$. Neglecting the transverse structure of $N(\xi)$, or the dependence of the photon distribution on the transverse variable $\vec{r}_\perp$, we can use

$$\vec{E}_\omega(\vec{r}) = A_\omega(x)\,e^{i\vec{k}_\perp \cdot \vec{r}_\perp}\vec{e}_\omega \qquad (5.46)$$

where $\vec{e}_\omega$ is the unit polarization vector and A_ω is the spectral field amplitude.

Equation (5.45) now reduces to

$$\frac{\partial^2 A_\omega}{\partial x^2} + k^2 A_\omega = -i\omega\mu_0(\vec{e}_\omega \cdot \vec{e}_x)\int_{-\infty}^\infty v_g Q_{ph}N(\xi)\,e^{i\omega t}\,dt \qquad (5.47)$$

where

$$k^2 = \frac{\omega^2}{c^2}\epsilon(\omega) - k_\perp^2. \qquad (5.48)$$

The solution of equation (5.47) for radiation emitted in the backward direction (or in the direction of negative values of x) is given by

$$A_\omega(x) = -\mu_0\frac{\omega}{2kc}(\vec{e}_\omega \cdot \vec{e}_x)\,e^{-ikx}\int_{-\infty}^\infty dt \int_{x_0}^x dx\,v_g Q_{ph}N(\xi)\,e^{ikx+i\omega t}. \qquad (5.49)$$

Notice that, due to the factor $(\vec{e}_\omega \cdot \vec{e}_x)$ we can only observe a radiated transverse field if $k_\perp$ is not equal to zero. Let us assume the case of radiation due to a modulation in the electron mean density

$$n_0(x) = n_0 + \tilde{n}\cos(\tilde{k}x). \qquad (5.50)$$

This leads to

$$k_p^2 = \frac{\omega_p^2}{v_g^2} \simeq \frac{\omega_{p0}^2}{v_{g0}^2}\left[1 + \frac{\tilde{n}}{n_0}\cos(\tilde{k}x)\right]. \qquad (5.51)$$

The modulations in the photon group velocity also exist, but they were neglected ($v_g \simeq v_{g0}$) because they would only give second-order contributions. Using this expression in equation (5.37) we can get from equation (5.49)

$$A_\omega(x) = A\,e^{-ikx}\int_{-\infty}^\infty dt \int_{x_0}^x dx\,N(\xi)\cos(\tilde{k}x)\,e^{ikx+i\omega t} \qquad (5.52)$$

with

$$A = \mu_0(\vec{e}_\omega \cdot \vec{e}_x)\frac{\omega}{\omega_0}\frac{e}{m}\frac{\omega_{p0}^2}{kcv_{g0}}\frac{\tilde{n}}{n_0}. \qquad (5.53)$$

The asymptotic field radiated very far away from the emitting region can then be written as

$$A_\omega(x \to -\infty) = A\,e^{-ikx}\int_{-\infty}^\infty dt \int_{-\infty}^\infty d\xi\,N(\xi)$$
$$\times \cos[\tilde{k}(\xi + v_{g0}t)]\,e^{ik(\xi+v_{g0}t)+i\omega t}. \qquad (5.54)$$

Integration in time leads to

$$A_\omega(x \to -\infty) = \pi A \int d\xi\, N(\xi) \Big[e^{i(\tilde{k}+k)\xi} \delta(\omega - \omega_1)$$
$$+ e^{-i(\tilde{k}-k)\xi} \delta(\omega - \omega_2) \Big] \tag{5.55}$$

with

$$\omega_1 = v_{g0}(\tilde{k} - k), \quad \omega_2 = v_{g0}(\tilde{k} + k). \tag{5.56}$$

This result shows that an electromagnetic wavepacket (or equivalently a photon bunch described by $N(\xi)$) travelling in a plasma with a modulated electron density, radiates in two characteristic frequencies which depend on the periodicity scale of the modulation. This is similar to the radiation of electrons travelling in a vacuum in the presence of a modulated magnetic field (an electron ondulator), showing once more the analogies between a photon in a plasma and a charged particle. Such a radiation process [69] can then be called a photon ondulator.

5.3 Photon transition radiation

Let us examine another example of a radiation process directly due to the existence of the equivalent charge: the possibility of transition radiation emitted by a photon bunch at a plasma boundary. This process is asociated with the sudden disappearence of the equivalent charge [71].

In order to simplify the problem, we will assume a nearly monochromatic burst of photons moving along the x-axis. We have then

$$N_k(\vec{r}, t) = (2\pi)^3 N(\xi, \vec{r}_\perp) \delta(\vec{k} - \vec{k}_0) \tag{5.57}$$

where $\xi = x - v_0 t$, and $v_0 = v_k$ is the group velocity for $\vec{k} = k_0 \vec{e}_x$.

Let us also assume a plasma–vacuum transition layer described by

$$\omega_p^2(x) = \omega_{p0}^2 f(x). \tag{5.58}$$

A plausible choice for the shape function is $f(x) = [1 + \tanh(x/\Delta)]/2$, where Δ is the width of the boundary layer. The nonlinear current $\vec{J}_{2NL}$ reduces to

$$\vec{J}_{2NL} = Q_0 N(\xi, \vec{r}_\perp) v_0(x) f(x) \vec{e}_x \tag{5.59}$$

with $Q_0 = Q_{ph}(k)$ for $k = k_0$ and for $\omega_p = \omega_{p0}$. We know that the group velocity v_0 is determined by

$$v_0(x) = c\sqrt{1 - (\omega_{p0}/\omega_0)^2 f(x)}. \tag{5.60}$$

In order to give a simple physical explanation of the photon transition radiation, we consider the particular form of a sharp plasma boundary. We consider the simplest case of a boundary with infinitesimal width $\Delta \rightarrow 0$, such that $f(x) = H(-x)$, where $H(x)$ is the Heaviside function. We also assume that the duration of the bunch of photons is negligible, so that we can write

$$N(\xi, \vec{r}_\perp) = N(\vec{r}_\perp)\delta(\xi). \tag{5.61}$$

In this case, the nonlinear current becomes

$$\vec{J}_{2NL} = -Q_0 N(\vec{r}_\perp)v_0 H(-x)\delta(x - v_0 t)\vec{e}_x. \tag{5.62}$$

This means that the source term in the wave equation can be written as

$$\frac{\partial}{\partial t}\vec{J}_{2NL} = Q_0 N(\vec{r}_\perp)v_0^2 H(-x)\delta'(x - v_0 t)\vec{e}_x$$
$$= -Q_0 N(\vec{r}_\perp)v_0\delta(t)\delta(x)\vec{e}_x. \tag{5.63}$$

We see that the source term is located both in space and time. In space, it is located precisely at the plasma boundary. In time it is located at the instant when the photon bunch crosses the boundary. This means that, in the vacuum region ($x > 0$), the transverse field radiated by the boundary is determined by

$$\left(\nabla^2 - \frac{1}{c^2}\frac{\partial^2}{\partial t^2}\right)\vec{E} = -\mu_0 Q_0 v_0 N(\vec{r}_\perp)\delta(t)\delta(x)\vec{e}_x. \tag{5.64}$$

This equation shows that an infinitely sharp boundary will radiate an infinitely large spectrum. This is very similar to the well-known case of bremsstrahlung radiation, where the particle velocity changes and its charge remains constant (because it is an invariant). In contrast here, the velocity of the radiating particle remains nearly constant and its equivalent charge goes to zero.

In order to obtain a finite spectral width, we can replace equation (5.61) by a more realistic description of a bunch of photons coming out of the plasma. For instance, we can assume a Gaussian bunch described by

$$N(\xi, \vec{r}_\perp) = \frac{N(\vec{r}_\perp)}{\sigma\sqrt{\pi}}\exp\left[-\frac{\xi^2}{\sigma^2}\right] \tag{5.65}$$

where σ is the bunch width. The nonlinear current becomes

$$\vec{J}_{2NL} = -Q_0 v_0 N(\vec{r}_\perp)H(-x)\frac{\vec{e}_x}{\sigma\sqrt{\pi}}\exp\left[-\frac{(x - v_0 t)^2}{\sigma^2}\right]. \tag{5.66}$$

The wave equation for the secondary radiated field can now be written as

$$\left(\nabla^2 - \frac{1}{c^2}\frac{\partial^2}{\partial t^2}\right)\vec{E}_2 = -\mu_0 Q_0 v_0 N(\vec{r}_\perp)\frac{\vec{e}_x}{\tau\sqrt{\pi}}\exp\left[-\frac{t^2}{\tau^2}\right]\delta(x) \tag{5.67}$$

where τ is here σ/v_0.

Following a procedure similar to that of the previous section, we arrive at a general solution that can be written in the following simple way:

$$A_\omega(z) = A_0(\omega)\, e^{ikx} \tag{5.68}$$

with

$$A_0(\omega) = i\frac{\mu_0}{2k}\frac{Q_0}{\sqrt{\tau}}\, v_0(\hat{e}_\omega^* \cdot \vec{e}_x) N(\vec{k}_\perp) \exp\left[-\frac{\omega^2\tau^2}{4}\right] \tag{5.69}$$

where $k^2 = (\omega/c)^2 - k_\perp^2$, and

$$N(\vec{k}_\perp) = \int N(\vec{r}_\perp)\, e^{-i\vec{k}_\perp \cdot \vec{r}_\perp}\, d\vec{r}_\perp. \tag{5.70}$$

This result shows that the energy radiated by the disappearing equivalent charge is proportional to the square of this charge, Q_0^2, and to the square of the photon velocity, v_0^2, as in the usual linear [57] and nonlinear [62] transition radiation processes. Furthermore, radiation is not emitted along the particle direction of motion because of the geometric factor $(\hat{e}_\omega^* \cdot \vec{e}_x)$. As already noticed, this factor is equal to zero for parallel radiation $\vec{k}_\perp = 0$, because of the transverse nature of the radiated field.

Finally, let us discuss the importance of the transverse dimensions of the burst of particles crossing the boundary. If these transverse dimensions are negligible, we can write $N(\vec{r}_\perp) = N_0\delta(\vec{r}_\perp)$. In this case, we can use $N(\vec{k}_\perp) = 1$ in equation (5.70), and the radiated field will have a maximum for nearly perpendicular direction of propagation, where $(\vec{e}_\omega^* \cdot \vec{e}_x) \simeq 1$. In the opposite limit of an infinitely large bunch, such that $N(\vec{r}_\perp) = N_0$, and $N(\vec{k}_\perp) = (2\pi)^2 N_0\delta(\vec{k}_\perp)$, we will have no transition radiation at all.

These simple calculations can obviously be refined, and the general case of an arbitrary photon distribution crossing an arbitrary plasma boundary can be studied with the aid of equations (5.37, 5.43). This shows that, in quite general conditions, a primary distribution of photons can radiate a large spectrum of electromagnetic waves (or secondary photons) when they cross a plasma boundary with nearly constant velocity. This is a direct consequence of their equivalent electric charge in a plasma.

This new kind of transition radiation is qualitatively similar to the usual transition radiation of charged particles (for instance electrons) moving across a dielectric discontinuity with constant velocity. In particular, the radiated power is proportional to the square of the particle velocity and of the particle equivalent charge. However, in some respects, the new type of transition radiation discussed here is also quite similar to the usual bremsstrahlung, with the difference that, instead of a change in velocity, we have a change in the value of the particle charge.

5.4   Photon Landau damping

In section 4.3 we considered the interaction of a photon distribution with an electron plasma wave moving with a relativistic phase velocity. But it can be easily recognized that our approach was not self-consistent in the sense that the electron plasma wave was defined *a priori*, and its evolution was assumed to be independent of the radiation spectrum. Here we return to the same problem, but with a more consistent view of the physics of the photon–plasma interaction, in the sense that the electron plasma oscillations will be coupled to the photon field [14].

In equilibrium, we can characterize the plasma by its electron mean density n_0, and the radiation spectrum by some distribution of photons, N_{k0}, for instance the Planck distribution for a given plasma temperature T. If the plasma is perturbed, the photon field will also be perturbed because they are coupled to each other.

This means that we can write, for the total electron plasma density n and for the total photon distribution N_k, the following expressions:

$$n = n_0 + \bar{n}, \quad N_k = N_{k0} + \bar{N}_k. \tag{5.71}$$

The objective of the present section is to show how the coupling between these two perturbed quantities can be described, and what are the new physical processes associated with it. If we use the electron fluid equations and Poisson's equation for the electrostatic field, we can derive the following equation for the electron density perturbation:

$$\frac{\partial^2 \bar{n}}{\partial t} + \omega_{p0}^2 \bar{n} - 3v_e^2 \nabla^2 \bar{n} = -n_0 \nabla \cdot \left\langle \frac{\partial \vec{v}}{\partial t} \right\rangle \tag{5.72}$$

where ω_{p0} is the unperturbed electron plasma frequency and $v_e = \sqrt{T/m}$ is the electron thermal velocity.

In this equation, the right-hand side describes the ponderomotive force due to the photon field. We can write it more explicitly as

$$\left\langle \frac{\partial \vec{v}}{\partial t} \right\rangle = -\frac{1}{2} \nabla |v|^2 = -\frac{1}{2} \left(\frac{e}{m} \right)^2 \nabla \int \frac{|E_k|^2}{\omega_k^2} \frac{d\vec{k}}{(2\pi)^3}. \tag{5.73}$$

Here we have assumed the linear solution for the electron motion in the radiation field $\vec{v} = -i(e\vec{E}_k/m\omega_k)$. This can also be written in terms of the photon number distribution, if we use the plane wave definition

$$N_k = \frac{\epsilon}{8\hbar} \left(\frac{\partial R}{\partial \omega} \right)_k |E_k|^2 \tag{5.74}$$

where $R \equiv R(\omega, k) = 0$ is the photon dispersion relation.

Replacing this in equation (5.72), we obtain

$$\frac{\partial^2 \tilde{n}}{\partial t} + \omega_{p0}^2 \tilde{n} - 3v_e^2 \nabla^2 \tilde{n} = 4\hbar \frac{\omega_{p0}^2}{m} \nabla^2 \int \frac{\tilde{N}_k}{\omega_k^2 (\partial R/\partial \omega)_k} \frac{d\vec{k}}{(2\pi)^3}. \tag{5.75}$$

In the absence of any radiation field, $\tilde{N}_k = 0$, the right-hand side of this equation is equal to zero and, for perturbations of the form $\tilde{n} \sim \exp \mathrm{i}(\vec{k} \cdot \vec{r} - \omega t)$, this equation will lead to the well-known linear dispersion relation for electron plasma waves

$$\omega^2 = \omega_{p0}^2 + 3v_e^2 k^2. \tag{5.76}$$

It should be kept in mind that, in this equation, the frequency ω and the wavenumber k are related to the electron plasma wave spectrum, and not to the photon spectrum. Furthermore, it is known from the plasma kinetic theory that this dispersion relation is only valid if the electron Landau damping is negligible, which implies that the phase velocity of the electrostatic wave has to be much larger than the electron thermal velocity: $(\omega/k) \gg v_e$. We will come back to that point later.

In the general case where the radiation field is present and we have $\tilde{N}_k \neq 0$, equation (5.75) is coupled with the kinetic equation for the photon field which, after an appropriate linearization, determines $\tilde{N}_k$ as a function of N_{k0} and $\tilde{n}$:

$$\frac{\partial \tilde{N}_k}{\partial t} + \vec{v}_k \cdot \frac{\partial \tilde{N}_k}{\partial \vec{r}} = -\vec{f} \cdot \frac{\partial N_{k0}}{\partial \vec{k}}. \tag{5.77}$$

According to our theory of photon acceleration, the force acting on the photons is determined by $\vec{f} = -\nabla \omega_k$, where $\omega_k = \sqrt{\omega_p^2 + k^2 c^2}$. This leads to the following explicit expression for the force acting on the photons:

$$\vec{f} = -\frac{1}{2\omega_k} \frac{e^2}{\epsilon_0 m} \nabla \tilde{n}. \tag{5.78}$$

The pair of equations (5.75, 5.77) will then describe the coupling between the electron plasma waves associated with the density perturbation $\tilde{n}$ and the photon number perturbations $\tilde{N}_k$. Let us assume sinusoidal perturbations of the form

$$\tilde{n}(\vec{r}, t) = \tilde{n}(t) \, e^{\mathrm{i}\vec{k}\cdot\vec{r}}, \quad \tilde{N}_{k'}(\vec{r}, t) = \tilde{N}_{k'}(t) \, e^{\mathrm{i}\vec{k}\cdot\vec{r}}. \tag{5.79}$$

Here, the value of the wavevector $\vec{k}$, characterizing the scale of the perturbation in both the electron density and the photon spectrum, should not be confused with the wavevector $\vec{k}'$ characterizing a specific Fourier component of the photon spectrum. In physical terms we are dealing with a situation which only makes sense when we have $\vec{k}' \gg \vec{k}$. This means that the electron plasma wave has a much larger scale length than the photons described by $\tilde{N}_{k'}$, which makes plausible the description of the transverse radiation field as a fluid of particles (the photons) evolving in a slowly varying background medium (the plasma and its electrostatic wave).

Using equation (5.79) in equations (5.75, 5.77), we obtain

$$\frac{\partial^2 \tilde{n}}{\partial t^2} + (\omega_{p0}^2 + 3k^2 v_e^2)\tilde{n} = -4\hbar k^2 \frac{\omega_{p0}^2}{m} \int \frac{\tilde{N}_{k'}}{\omega'^2 (\partial R/\partial \omega)_{\omega'}} \frac{d\vec{k}'}{(2\pi)^3} \qquad (5.80)$$

and

$$\frac{\partial \tilde{N}_{k'}}{\partial t} + i\vec{k} \cdot \vec{v}_{k'} \tilde{N}_{k'} = \frac{i}{2\omega'} \frac{e^2}{\epsilon_0 m} \tilde{n}\vec{k} \cdot \frac{\partial N_{k'0}}{\partial \vec{k}'} \qquad (5.81)$$

where $\omega' \equiv \omega(\vec{k}')$.

We can now follow the usual Landau approach [82], but apply it to the photon distribution and not to the electron distribution. This implies the use of a time Laplace transformation

$$\tilde{n}(\omega) = \int_0^\infty \tilde{n} \, e^{i\omega t} \, dt \qquad (5.82)$$

and a similar transformation for $\tilde{N}_{k'}(\omega)$, where ω is here a complex quantity.

From equation (5.81) we then get a relation between $\tilde{n}(\omega)$ and $\tilde{N}_{k'}(\omega)$:

$$\tilde{N}_{k'} = -\frac{\tilde{n}}{2n_0} \frac{\omega_{p0}^2}{\omega'} \frac{\vec{k} \cdot (\partial N_{k'0}/\partial \vec{k}')}{\omega - \vec{k} \cdot \vec{v}(\vec{k}')}. \qquad (5.83)$$

Similarly, from equation (5.80) we get

$$\left(\omega_{p0}^2 + 3k^2 v_e^2 - \omega^2\right)\tilde{n}$$

$$= 2\omega_{p0}^4 \frac{\hbar k^2}{m} \frac{\tilde{n}}{n_0} \int \frac{1}{\omega'^3 (\partial R/\partial \omega)_{\omega'}} \frac{\vec{k} \cdot (\partial N_{k'0}/\partial \vec{k}')}{\omega - \vec{k} \cdot \vec{v}(\vec{k}')} \frac{d\vec{k}'}{(2\pi)^3}. \qquad (5.84)$$

We can use $(\partial R/\partial \omega)_{\omega'} \simeq 2/\omega'$. From these two equations we can then derive the following expression:

$$\omega^2 = 3k^2 v_e^2 + \omega_{p0}^2 \left[1 - \frac{\hbar k^2}{mn_0}\omega_{p0}^2 \int \frac{1}{\omega'^2} \frac{\vec{k} \cdot (\partial N_{k'0}/\partial \vec{k}')}{\omega - \vec{k} \cdot \vec{v}(\vec{k}')} \frac{d\vec{k}'}{(2\pi)^3}\right]. \qquad (5.85)$$

In order to develop this integral we have to consider separately the parallel and the perpendicular photon motion with respect to the propagation of the electron plasma wave. This can be done by defining

$$\vec{k}' = p\frac{\vec{k}}{k} + \vec{k}'_\perp, \qquad \vec{v}(\vec{k}') = u(p, \vec{k}'_\perp)\frac{\vec{k}}{k} + \vec{v}_\perp. \qquad (5.86)$$

The integral in equation (5.85) becomes

$$\int \frac{1}{\omega'^2} \frac{\vec{k} \cdot (\partial N_{k'0}/\partial \vec{k}')}{\omega - \vec{k} \cdot \vec{v}(\vec{k}')} \frac{d\vec{k}'}{(2\pi)^3} = -\int \frac{d\vec{k}'_\perp}{(2\pi)^3} \int \frac{dp}{\omega'^2} \frac{\partial N_{k'0}/\partial p}{u - (\omega/k)}. \qquad (5.87)$$

Developing the parallel photon velocity u around the resonant value defined by $u(p_0) = (\omega/k)$

$$u(p, \vec{k}'_\perp) \simeq u(p_0, \vec{k}'_\perp) + (p - p_0) \left(\frac{\partial u}{\partial p} \right)_{p_0} \tag{5.88}$$

we get

$$\int \frac{dp}{\omega'^2} \frac{\partial N_{k'0}/\partial p}{u - (\omega/k)} \simeq \frac{1}{(\partial u/\partial p)_{p_0}} \int \frac{dp}{\omega'^2} \frac{\partial N_{k'0}/\partial p}{p - p_0}. \tag{5.89}$$

This last integral can be written in the standard form

$$I(z) \equiv \int \frac{h(z)}{z - z_0} \, dz = P \int \frac{h(z)}{z - z_0} \, dz + i\pi h(z_0) \tag{5.90}$$

where $P \int$ means the Cauchi principal part of the integral.

Replacing this result in equation (5.85) and using $\omega = \omega_r + i\gamma$, we get from the real part of the resulting equation

$$\omega_r^2 = 3k^2 v_e^2 \omega_{p0}^2 \left[1 + \frac{\hbar k^2}{mn_0} \omega_{p0}^2 P \int \frac{1}{p - p_0} \frac{\partial G_p}{\partial p} \, dp \right], \tag{5.91}$$

and from the imaginary part

$$\gamma = \pi \frac{\hbar k^2 \omega_{p0}^3}{2mn_0} \left(\frac{\partial G_p}{\partial p} \right)_{p_0}. \tag{5.92}$$

In these two expressions, G_p is a kind of reduced distribution function for the photon gas, which resulted from the integration of the equilibrium photon distribution $N_{k'0}$ over the perpendicular directions:

$$G_p = \int \frac{1}{\omega'^2(p_0)} \frac{N_{k'0}}{(\partial u/\partial p)_{p_0}} \frac{d\vec{k}'_\perp}{(2\pi)^3}. \tag{5.93}$$

Equations (5.91, 5.92) are the main results of this section. The first one represents the dispersion relation of electron plasma waves (in the high phase velocity regime) in the presence of a photon distribution. The second one determines the damping of these plasma waves due to the resonant interaction with the photon spectrum. Due to the similarities of this expression with the well-known electron Landau damping, this new damping effect can be called the *photon Landau damping*.

The similarities between the electron and the photon contributions can be stated in a more appropriate way if we notice that the dispersion relation (5.91) can be rewritten as

$$1 + \chi_e + \chi_{ph} = 0 \tag{5.94}$$

where χ_e is the electron susceptibility and χ_{ph} is the photon susceptibility.

According to equation (5.91) the electron susceptibility is

$$\chi_e = -\frac{\omega_{p0}^2}{\omega_r^2} - 3\frac{k^2 v_e^2}{\omega_r^2} \tag{5.95}$$

and the photon susceptibility is

$$\chi_{ph} = -\frac{\hbar k^2}{mn_0}\omega_{p0}^2 P \int \frac{1}{p - p_0}\frac{\partial G_p}{\partial p}\, dp. \tag{5.96}$$

As we have noticed already, the above expression for the electron susceptibility is only valid in the limit of very high phase velocities. In the general case,
χ_e would depend on the actual electron distribution function and moreover, an
electron Landau damping would have to be added to the above photon Landau
damping. These electron kinetic effects are well documented in plasma physics
textbooks and will not be described here.

The important point to notice here is that these electron kinetic effects are
negligible for electron plasma waves with high phase velocities, like those produced by intense laser pulses propagating in a plasma. In this case the resulting
electron plasma waves (or wakefields) have relativistic phase velocities nearly
equal to the laser pulse group velocity from which they originate.

The important conclusion of the above calculation is that, even in the absence of resonant electron populations which could exchange efficiently their
energy with the electron plasma wave, the relativistic plasma perturbations can
nevertheless exchange energy with the photon gas. The resulting wave damping
is described by equation (5.92). This means that photon Landau damping can
replace the electron Landau damping, showing that in many respects the photon
dynamics in the field of an electrostatic wave in a plasma is similar to the electron
dynamics, as already documented in chapter 3 for single photon trajectories.

For a plasma in thermal equilibrium at a temperature T, the Planck distribution

$$N_{k'0} = \frac{\omega'^2}{\pi^2 c^3}\frac{1}{\exp(\hbar\omega'/T) - 1} \tag{5.97}$$

where $\omega' = \sqrt{\omega_{p0}^2 + k'^2 c^2}$ should be used in the above equations.

Because of the derivative contained in the expression for the photon Landau
damping (5.92) this equilibrium distribution will always lead to a negative value
of γ, or to a real wave damping. However, the situation can change if a photon
beam is superposed on the Planck distribution. Then, in some regions of phase
velocities, the derivative can be positive, indicating the possibility of a wave
instability or wave growth. This will be illustrated below.

It should also be noticed that a quasi-linear diffusion coefficient for the photons can be derived from this theory, showing how the photon spectrum changes
in the presence not of a single plasma wave as before, but in the presence of a
spectrum of electron plasma waves [14].

5.5 Photon beam plasma instabilities

In order to illustrate the new physical aspects contained in the dispersion relation (5.94, 5.96) let us consider a one-dimensional problem where

$$N_{k'0} = (2\pi)^2 N_{k'} \delta(\vec{k}'_\perp). \tag{5.98}$$

The photon susceptibility will then reduce to

$$\chi_{\text{ph}} = \frac{\hbar k^2}{mn_0} \frac{\omega_{p0}^2}{\omega^2} \int \frac{1}{\omega_{k'}^2} \frac{\partial N_{k'}/\partial k'}{(\omega/k) - v(k')} \frac{dk'}{2\pi} \tag{5.99}$$

where $k' \equiv k'_\parallel$. Using the photon (group) velocity

$$v(k') = \frac{c^2 k'}{\omega_{k'}} \tag{5.100}$$

we obtain, after integration by parts,

$$\int \frac{1}{\omega_{k'}} \frac{\partial N_{k'}/\partial k'}{(\omega/k) - (c^2 k'/\omega_{k'})} \frac{dk'}{2\pi}$$
$$= -\frac{1}{c^2} \int \frac{1}{\omega_{k'}} \frac{N_{k'}}{[(\omega/k)(\omega_{k'}/c^2) - k']^2} \frac{dk'}{2\pi}. \tag{5.101}$$

Let us assume a mono-energetic photon beam propagating across the plasma:

$$N_{k'} = 2\pi N_0 \delta(k' - k'_0). \tag{5.102}$$

Using $\omega'_0 = \omega_{k'_0}$ we can write, after performing the integration,

$$\chi_{\text{ph}} = -\frac{\hbar k^4}{mn_0} \frac{\omega_{p0}^4}{\omega^2 \omega'_0} \frac{k^2 c^2}{\omega'_0} \frac{N_0}{(\omega - k u_0)^2} \tag{5.103}$$

where $u_0 = v(k'_0)$.

Furthermore, if we neglect the electron thermal effects, the electron susceptibility reduces to

$$\chi_e = -\frac{\omega_{p0}^2}{\omega^2}. \tag{5.104}$$

The dispersion relation for the electron plasma waves (5.94) can then be written as

$$1 - \frac{\omega_{p0}^2}{\omega^2} \left[1 + \frac{\Omega^2}{(\omega - k u_0)^2} \right] = 0 \tag{5.105}$$

where we have used

$$\Omega^2 = \frac{\hbar k^4 c^2}{mn_0} \frac{\omega_{p0}^2}{\omega_0'^3} N_0. \tag{5.106}$$

We can see from equation (5.105) that this quantity Ω plays nearly the same role in the photon susceptilibity that is played by the electron plasma frequency ω_{p0} in the electron susceptibility. Due to that, we can call it the effective photon plasma frequency.

But the analogy with a plasma frequency can be explored even further, and we can define a photon plasma frequency Ω_{ph}, formally identical to the electron plasma frequency, such that

$$\Omega_{ph}^2 = \frac{Q_{ph}^2 N_0}{\epsilon_0 m_{eff}} \tag{5.107}$$

where Q_{ph} is the photon equivalent charge as determined by equation (5.37), and $m_{eff} = \hbar\omega_{p0}/c^2$ is the photon effective mass introduced in chapter 2. This means that we can write

$$\Omega_{ph}^2 = \frac{4\hbar N_0}{mn_0} \frac{\omega_{p0}^5}{(\omega_0' c)^2}. \tag{5.108}$$

Comparing this with equation (5.106), we obtain

$$\Omega^2 = \frac{1}{4} \left(\frac{kc}{\omega_{p0}} \right)^4 \frac{\omega_{p0}}{\omega_0'} \Omega_{ph}^2. \tag{5.109}$$

Noting that, for the electron plasma waves resonant with the photon beam, we always have $k^2 c^2 \simeq \Omega_{p0}^2 \simeq \omega^2$, we can also write the approximate expression

$$\Omega^2 \simeq \frac{\omega_{p0}}{4\omega_0'} \Omega_{ph}^2. \tag{5.110}$$

We see that the effective photon plasma frequency Ω is always significantly larger than the photon plasma frequency Ω_{ph} as defined by equation (5.107). But, the fact that this quantity Ω_{ph} appears in the photon susceptibility shows that the concept of the photon equivalent charge Q_{ph} is already implicit in the photon susceptibility term and in the photon Landau damping. If the photon effective charge did not exist, the photon Landau damping would not be possible. We can then conclude that the high-frequency photons behave in a plasma like any other charged particle.

Coming back to the dispersion relation (5.105), we can assume the resonant condition $\omega_{p0} = ku_0$, and we can write $\omega = \omega_{p0} + i\mu$. We can then obtain the growth rate for the photon beam plasma instability as

$$\mu = \frac{\sqrt{3}}{2} \omega_{p0} \left(\frac{\Omega^2}{2\omega_{p0}^2} \right)^{1/3}. \tag{5.111}$$

We notice that this growth rate is proportional to the power $1/3$ of the density of photons in the beam:

$$\mu \propto N_0^{1/3}. \tag{5.112}$$

This instability is comparable to the usual beam plasma instability produced by the interaction of a monoenergetic electron beam (with mean velocity u_0) with a plasma, for which the growth rate is also proportional to the density of particles in the beam to the power $1/3$. This shows again that the photons behave, in this respect, just like electrons.

The photon beam plasma instability should also be compared with the forward Raman scattering [50], which occurs when a photon beam (with a well-defined field phase) interacts with a plasma. Electron plasma waves are also produced here, but the physical mechanism is completely different because it implies that the usual resonant conditions for wave decay are satisfied: $\omega_1' = \omega_0' \pm \omega$ and $k_1' = k_0' \pm k$, where ω_1' and k_1' are the frequency and wavenumber of the scattered photons. This corresponds to the well-known energy and momentum conservation rules, which are clearly distinct from the resonance condition $\omega = ku_0$, appling to our photon beam plasma instability.

Furthermore, the maximum growth rate for the forward Raman scattering is determined by

$$\mu \simeq \frac{\omega_{p0}^2}{2\sqrt{2}\omega_0'} \frac{v_{os}}{c} \tag{5.113}$$

where $v_{os} = eE_0/m\omega_0'$, and E_0 is the amplitude of the electric field associated with the incident photon beam.

We see that this new growth rate is proportional to $N_0^{1/2}$, in contrast with the result of equation (5.112), thus showing that the photon beam plasma instability is a new kind of instability, which is clearly distinct from the usual Raman decay instability.

5.6 Equivalent dipole in an optical fibre

We have shown that it is possible to define an equivalent photon charge in a plasma, which is a different way of describing the ponderomotive force or radiation pressure effects. Here we examine a similar concept for non-ionized dielectric media.

In particular, we examine a laser pulse propagation along an optical fibre having a second-order nonlinearity and show that an equivalent electric dipole can be derived [70]. The difference with respect to the plasma case is related to the absence of free electrons in this medium.

From Maxwell's equations, in the absence of charge and current distributions, we can establish the following equation for the propagating electric field:

$$\nabla^2 \vec{E} - \nabla \left(\nabla \cdot \vec{E} \right) - \frac{1}{c^2} \frac{\partial^2 \vec{E}}{\partial t^2} = \mu_0 \frac{\partial^2 \vec{P}}{\partial t^2}. \tag{5.114}$$

The polarization vector is $\vec{P} = \vec{P}_L + \vec{P}_{NL}$, with the linear part defined by

$$\vec{P}_L(t) = \epsilon_0 \int_0^\infty \chi^{(1)}(\tau) \cdot \vec{E}(t - \tau) \, d\tau. \tag{5.115}$$

The nonlinear part of the polarization vector, if we neglect dispersion effects, can be written as

$$\vec{P}_{NL}(t) = \epsilon_0 \chi^{(2)} \cdot \vec{E}(t)\vec{E}(t) + \epsilon_0 \chi^{(3)} \cdot \vec{E}(t)\vec{E}(t)\vec{E}(t). \tag{5.116}$$

We now assume, as we already did for the plasma case, that the electric field can be divided into two terms: the dominant field associated with the propagating pulse $\vec{E}_p$, and a secondary perturbation or radiation field, $\vec{E}_r$, such that $|E_r| \ll |E_p|$:

$$\vec{E} = \vec{E}_p + \vec{E}_r. \tag{5.117}$$

To the lowest order in the perturbation field $\vec{E}_r$, we can write from equations (5.114–5.117)

$$\nabla^2 \vec{E}_p - \nabla \left(\nabla \cdot \vec{E}_p \right) - \frac{1}{c^2} \frac{\partial^2}{\partial t^2} \left[\vec{E}_p + \int_0^\infty \chi^{(1)}(\tau) \cdot \vec{E}(t - \tau) \, d\tau \right]$$
$$= \frac{1}{c^2} \frac{\partial^2}{\partial t^2} \chi^{(3)} \cdot |E_p|^2 \vec{E}_p. \tag{5.118}$$

From this equation we can obtain linear (if $\chi^{(3)} = 0$) or nonlinear (if $\chi^{(3)} \neq 0$) wave pulse solutions. In general, for propagation along an optical fibre, we can write both types of solution in the general form

$$\vec{E}_p(\vec{r}, t) = \frac{\hat{a}_p}{2} \left[R(\vec{r}_\perp) A(z, t) \, e^{i(kz - \omega t)} + \text{c.c.} \right]. \tag{5.119}$$

This field is assumed to propagate along the fibre axial directon z with negligible pulse shape deformation, and with a group velocity v_g.

To the first order in the radiation field $\vec{E}_r$, and assuming for convenience that $\chi^{(2)}$ is of the same order of $\vec{E}_r$, we get from equations (5.114–5.117)

$$\nabla^2 \vec{E}_r - \nabla \left(\nabla \cdot \vec{E}_r \right) - \frac{1}{c^2} \frac{\partial^2}{\partial t^2} \left[\vec{E}_r + \int_0^\infty \chi^{(1)} \cdot \vec{E}_r(t - \tau) \, d\tau \right]$$
$$= \mu_0 \frac{\partial}{\partial t} \vec{J}_r \tag{5.120}$$

where $\vec{J}_r$ is the nonlinear polarization current, defined by

$$\vec{J}_r = \epsilon_0 \frac{\partial}{\partial t} \left[\chi^{(2)} \cdot \vec{E}_p \vec{E}_p + \chi^{(3)} \cdot |E_p|^2 \hat{a}_p \hat{a}_p \vec{E}_r \right]. \tag{5.121}$$

Using equation (5.119), we can see that

$$\vec{E}_p(\vec{r}, t)\vec{E}_p(\vec{r}, t) = \frac{1}{4}\hat{a}_p\hat{a}_p\left[R^2(\vec{r}_\perp)A^2(z, t)e^{2i(kz-\omega t)}\right.$$

$$\left. + |R(\vec{r}_\perp)|^2|a(z, t)|^2 + \text{c.c.}\right]. \tag{5.122}$$

The first term oscillates in time as $\exp(2i\omega t)$ and is associated with harmonic generation. Its physical content is well understood and, for this reason, we can ignore it here.

The term in $\chi^{(3)}$ describes the scattering of a probe radiation field by the main pulse and can also be ignored because it does not play any role in the effect to be discussed here. For simplicity, we will use $\chi^{(2)} \cdot \hat{a}_p\hat{a}_p = \chi^{(2)}\hat{e}$ in the remaining term.

The nonlinear polarization current $\vec{J}_r$ is then reduced to

$$\vec{J}_r = \frac{\epsilon_0}{2}\chi^{(2)}\frac{\partial}{\partial t}W(\vec{r}, t)\hat{e} \tag{5.123}$$

where we have used

$$W(\vec{r}, t) = |R(\vec{r}_\perp)|^2|A(z, t)|^2. \tag{5.124}$$

It is obvious that, if the main pulse moves along the fibre with no significant change in its envelope shape, we can write

$$W(\vec{r}, t) \equiv W(\vec{r}_\perp, z - v_g t). \tag{5.125}$$

This means that the time derivative appearing in equation (5.123) can be replaced by a space derivative

$$\frac{\partial}{\partial t}W = -v_g\frac{\partial}{\partial z}W. \tag{5.126}$$

As a result of this replacement, we can rewrite the nonlinear polarization current as the product of a charge with a velocity

$$\vec{J}_r = v_g Q(\vec{r}, t)\hat{e}. \tag{5.127}$$

This quantity $Q(\vec{r}, t)$ can be called the equivalent charge distribution of the main pulse, and it is determined by

$$Q(\vec{r}, t) = -\frac{\epsilon_0}{2}\chi^{(2)}\frac{\partial}{\partial z}W(\vec{r}_\perp, z - v_g t). \tag{5.128}$$

This is clearly a dipole charge distribution because the space derivative reverses its sign when we move from the front to the rear of the pulse. This clearly contrasts with the plasma case where, due to the existence of free electrons which are pushed away by the electromagnetic pulse, the equivalent charge distribution reduces to a negative monopole charge.

An interesting feature is that the total equivalent charge here is equal to zero, as can be seen by integrating the distribution (5.128) along the pulse distribution:

$$\int_{-\infty}^{\infty} Q(z,t)\,dz = -\frac{\epsilon_0}{2}\chi^{(2)}\int \frac{\partial}{\partial z}W\,dz = -\frac{\epsilon_0}{2}\chi^{(2)}[W]_{-\infty}^{\infty} = 0. \qquad (5.129)$$

This means that we have here total charge conservation, which is physically quite reassuring. Furthermore, this property is independent of the actual shape of the pulse.

Let us consider the interesting limiting case of a rectangular pulse: the equivalent charge distribution reduces to two point charges, one negative and the other positive, corresponding to

$$Q(\vec{r},t) = -Q_{eq}(\vec{r}_\perp)\left[\delta\left(z-\frac{L}{2}-v_g t\right) - \delta\left(z+\frac{L}{2}-v_g t\right)\right] \qquad (5.130)$$

where

$$Q_{eq} = \frac{\epsilon}{2}\chi^{(2)}W_0(\vec{r}_\perp) \qquad (5.131)$$

for a pulse of length L and maximum intensity W_0.

Replacing equation (5.127) in equation (5.120) we can determine the secondary field $\vec{E}_r$ created by this moving dipole:

$$\nabla^2\vec{E}_r - \nabla\left(\nabla\cdot\vec{E}_r\right) - \frac{1}{c^2}\frac{\partial^2}{\partial t^2}\left[\vec{E}_r + \int_0^{\infty}\chi^{(1)}\cdot\vec{E}_r(t-\tau)\,d\tau\right]$$

$$= \frac{1}{c^2}\frac{\partial}{\partial t}v_g Q(\vec{r},t)\hat{e}. \qquad (5.132)$$

It should be noticed that this secondary field is not always a radiation field. For instance, in a homogeneous medium, where both the main pulse velocity v_g and the charge amplitude Q_{eq} are constant, it will reduce to a near field. But even in this case we can imagine an external loop where the existence of such a field can eventually be detected.

Radiation will typically occur in two distinct situations: first, when the group velocity v_g changes due to a space dependence of the linear susceptibility of the medium $\chi^{(1)}$; second, when the charge amplitude changes due to the space variation of the nonlinear susceptibility $\chi^{(2)}$.

These two radiation mechanisms are clearly distinct from those usually considered in the literature [2], which are mainly concerned with soliton propagation in inhomogeneous but centro-symmetric media, where the second-order nonlinearities are forbidden ($\chi^{(2)} = 0$). In this case, the source of radiation is a linear term of the form $\delta\chi^{(1)}E_p$, where $\delta\chi^{(1)}$ is the perturbation of the linear susceptibility due to the inhomogeneities of the medium. In contrast with our present radiation mechanism, which is clearly associated with an acceleration of the moving dipole, these more common aspects of secondary radiation can be

(a)

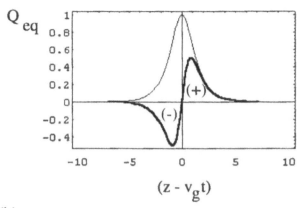

$$(z - v_g t)$$

(b)

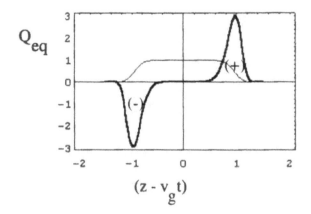

$$(z - v_g t)$$

Figure 5.1. Equivalent dipole distribution for (a) a sech laser pulse, and (b) a nearly rectangular pulse.

described as scattering of the lowest order pulse field by the local inhomogeneities of the medium.

Let us study equation (5.132) in more detail, and introduce a time Fourier transformation of the field as

$$\vec{E}_r(\vec{r}, t) = \int_{-\infty}^{\infty} \vec{E}_\omega(\vec{r}) \, e^{-i\omega t} \frac{d\omega}{2\pi}. \tag{5.133}$$

The Fourier components $\vec{E}_\omega$ will be determined by the equation

$$\left[\nabla^2 + \frac{\omega^2}{c^2}\epsilon(\omega)\right]\vec{E}_\omega - \nabla\left(\nabla \cdot \vec{E}_\omega\right) = i\frac{\omega}{c}\hat{e} \int_{-\infty}^{\infty} v_g(z)Q(\vec{r}_\perp, z, t)\,e^{i\omega t}\,dt$$

(5.134)

where $\epsilon(\omega)$ is the linear dielectric function of the medium.

Let us retain our attention on radiated transverse fields, such that $\nabla \cdot \vec{E}_\omega = 0$. Furthermore, in a cylindric geometry, we can use

$$\vec{E}_\omega(\vec{r}) = R_\omega(\vec{r}_\perp)A_\omega(z)\hat{e}_\omega.$$

(5.135)

Replacing it in the above equation, we get

$$\left[A_\omega \nabla_\perp^2 R_\omega + R_\omega \frac{\partial^2}{\partial z^2}A_\omega\right] + \frac{\omega^2}{c^2}\epsilon(\omega)R_\omega A_\omega$$

$$= i\frac{\omega}{c}(\hat{e} \cdot \hat{e}_\omega)v_g(z)\int_{-\infty}^{\infty} Q(\vec{r}, t)\,e^{i\omega t}\,dt.$$

(5.136)

This equation can be solved by considering first the corresponding homogeneous equation

$$\nabla_\perp^2 R_\omega + \frac{\partial^2}{\partial z^2}A_\omega + \frac{\omega^2}{c^2}\epsilon(\omega) = 0.$$

(5.137)

If we assume that the radial field amplitude R_ω is determined by the equation

$$\nabla_\perp^2 R_\omega = -k_\perp^2 R_\omega$$

(5.138)

where $k_\perp$ is the perpendicular wavenumber, the axial field amplitude A_ω will have the solution

$$A_\omega = A_\pm\,e^{\pm ikz}.$$

(5.139)

The axial wavenumber k is determined by the relation

$$k^2 = \frac{\omega^2}{c^2}\epsilon(\omega) - k_\perp^2.$$

(5.140)

Returning to equation (5.136), we can write the inhomogeneous equation for the axial field amplitude A_ω in the form

$$\frac{\partial^2 A_\omega}{\partial z^2} + k^2 A_\omega = i\frac{\omega}{c}(\hat{e} \cdot \hat{e}_\omega)v_g(z)\int_{-\infty}^{\infty} q(z, t)\,e^{i\omega t}\,dt$$

(5.141)

where $q(z, t)$ is the axial equivalent charge distribution, determined by

$$q(z, t) = \frac{\int R_\omega^*(\vec{r}_\perp)Q(\vec{r}_\perp, z, t)\,d\vec{r}_\perp}{\int |R_\omega(\vec{r}_\perp)|^2\,d\vec{r}_\perp}.$$

(5.142)

We now use the well-known method of the variation of parameters to solve equation (5.141). For forward propagation, we obtain

$$A_\omega(z)_+ = \frac{i}{w}\frac{\omega}{c}(\hat{e}\cdot\hat{e}_\omega)\,e^{ikz}\int dt\,e^{i\omega t}\int_{z_0}^z dz'\,v_g(z')g(z,t)\,e^{-ikz} \qquad (5.143)$$

where the Wronskian is $w = 2ik$.

To be more specific, let us assume that we have an axial modulation of the nonlinear properties of the optical fibre, which can be described by

$$v_g(z) = v_0 = \text{const}, \quad \chi^{(2)} = \chi_{20}\,[1 + \delta\cos(k_0 z)]. \qquad (5.144)$$

This means that we can replace, in the above solution, the axial equivalent charge by

$$q(z, t) = q_0(z, t)\delta\cos k_0 z. \qquad (5.145)$$

The forward field solution becomes

$$A_\omega(z)_+ = A_0\,e^{ikz}\int dt\,e^{i\omega t}\int_{z_0}^z dz'\,q_0(z', t)\cos(k_0 z)\,e^{-ikz} \qquad (5.146)$$

with an amplitude

$$A_0 = \frac{\omega}{2kc}(\hat{e}\cdot\hat{e}_\omega)v_0\delta. \qquad (5.147)$$

In the particular case of a rectangular pulse shape, as described by equation (5.130), we can write

$$q_0(z, t) = q_0\left[\delta\left(z - \frac{L}{2} - v_0 t\right) - \delta\left(z + \frac{L}{2} - v_0 t\right)\right]. \qquad (5.148)$$

We can now easily integrate equation (5.146) in z, for a very long fibre, and we obtain

$$A_\omega(z) = A_0 q_0\,e^{ikz}\int e^{i\omega t}\left[\cos\left(k_0\frac{L}{2} + k_0 v_0 t\right) e^{-ik(L/2)-ikv_0 t}\right.$$
$$\left. - \cos\left(k_0\frac{L}{2} - k_0 v_0 t\right) e^{ik(L/2)-ikv_0 t}\right] dt. \qquad (5.149)$$

From this equation it can then be concluded that two distinct frequencies are radiated by the main pulse when it propagates in a modulated optical fibre:

$$A_\omega(z) = 2\pi i A_0 q_0\left\{\sin\left[(k_0 - k)\frac{L}{2}\right]\delta(\omega - \omega_1)\right.$$
$$\left. + \sin\left[(k_0 + k)\frac{L}{2}\right]\delta(\omega - \omega_2)\right\} e^{ikz}. \qquad (5.150)$$

The values of the two frequencies are determined by

$$\omega_1 = v_0(k - k_0), \quad \omega_2 = v_0(k + k_0). \qquad (5.151)$$

When the wavelength of the radiated field is nearly equal to the characteristic wavelength of the fibre inhomogeneity, $k \simeq k_0$, we have $\omega_1 \simeq 0$ and $\omega_2 \simeq 2v_0k_0$. This means that, in this case, ω_2 is a kind of second harmonic.

Finally, we note that, for the more general situation of a pulse with an arbitrary shape, the solution (5.150) is replaced by

$$
\begin{aligned}
A_\omega(z)_+ = \pi A_0\, e^{ikz} \int q_0(\xi) \Big[& e^{i(k_0-k)\xi} \delta(\omega - \omega_1) \\
& + e^{-i(k_0-k)\xi} \delta(\omega - \omega_2) \Big]\, d\xi
\end{aligned}
\tag{5.152}
$$

with $\xi = x - v_0 t$.

This radiation process is a direct consequence of the existence of an equivalent dipole charge of the optical pulse propagating in an optical medium with second-order nonlinearities. As already noticed for the plasma case treated in section 5.2, it can be considered as the analogue of the electron ondulator, or another version of the photon ondulator.

Chapter 6

Full wave theory

In previous chapters we were able to describe a large number of different physical phenomena using the Hamiltonian approach to photon dynamics, either in its simple form of single photon trajectories or in its more elaborated version of the photon kinetic theory. However, the geometric optics approximation associated with this elegant and powerful approach is not capable of explaining such basic and sometimes important phenomena as partial reflection and mode coupling.

A full wave description of the electromagnetic radiation is therefore necessary, and it will be the subject of the present and the next chapters. Several papers have been devoted to this problem [26, 27, 78, 118, 123].

In chapter 3 we showed that an effect similar to the usual refraction, which we called time refraction, could take place. Here we will see that a new effect, which, in the same spirit, can be called time reflection, can also take place.

6.1   Space and time reflection

For comparison with the cases of time-varying media, we start first by reminding ourselves of the well-known derivation of the Fresnel formulae, by considering reflection and transmission of an incident electromagnetic wave at the boundary between two stationary dielectric media.

6.1.1   Reflection and refraction

Let us assume two different media, with dielectric constants ϵ_1 and ϵ_2, with a sharp boundary at the plane $x = 0$. If a plane wave with frequency ω and wavevector $\vec{k}_i$ propagates along the x-axis and interacts with this boundary, the total electric field will be determined by

$$\vec{E}(x,t) = \begin{cases} \vec{E}_i \exp i(k_i x - \omega t) + \vec{E}_r \exp i(k_r x - \omega t), & (x < 0) \\ \vec{E}_t \exp i(k_t x - \omega t), & (x > 0). \end{cases} \quad (6.1)$$

123

This field has three different terms, corresponding to the incident, the reflected and the transmitted waves. These waves have to satisfy the dispersion relations of both media, which implies that

$$k_i = \frac{\omega}{c} n_1, \quad k_r = -k_i, \quad k_t = k_i \frac{n_2}{n_1} \tag{6.2}$$

with the refractive indices $n_{1,2} = \sqrt{\epsilon_{1,2}}$.

In order to establish a relation between the field amplitudes $\vec{E}_i$, $\vec{E}_r$ and $\vec{E}_t$, we have to state the associated boundary conditions. These are determined by Maxwell's equations. In the absence of charge and current distributions in the media ($\rho = 0$, $\vec{J} = 0$), these equations are

$$\nabla \times \vec{E} = -\frac{\partial \vec{B}}{\partial t}, \quad \nabla \times \vec{H} = \frac{\partial \vec{D}}{\partial t} \tag{6.3}$$

and

$$\nabla \cdot \vec{D} = 0, \quad \nabla \cdot \vec{B} = 0. \tag{6.4}$$

It is well known that, in order to satisfy the first pair of equations, the components of $\vec{E}$ and $\vec{H}$ tangent to the boundary have to be continuous across this boundary:

$$(\vec{E}_1 - \vec{E}_2) \times \vec{e}_x = 0, \quad (\vec{H}_1 - \vec{H}_2) \times \vec{e}_x = 0. \tag{6.5}$$

From equations (6.3) we realize that, for plane waves propagating in non-magnetic media, we can write

$$\vec{B} \equiv \mu_0 \vec{H} = \frac{\vec{k} \times \vec{E}}{\omega}. \tag{6.6}$$

Assuming perpendicular polarization ($\vec{E} \cdot \vec{e}_x = 0$), and using the wavenumber relations (6.2), we can write the boundary conditions (6.5) in the form

$$E_i + E_r = E_t, \quad E_i - E_r = E_t \frac{n_2}{n_1}. \tag{6.7}$$

From this we derive the well-known Fresnel formulae for normal wave incidence

$$T \equiv \frac{E_t}{E_i} = \frac{2n_1}{n_1 + n_2}, \quad R \equiv \frac{E_r}{E_i} = \frac{n_1 - n_2}{n_1 + n_2}. \tag{6.8}$$

These expressions verify the simple relation

$$1 + R = T. \tag{6.9}$$

These results are extremely well known and we should not insist on their physical significance. We should instead use them as a reference guide for the less understood time-varying processes.

6.1.2 Time reflection

Let us turn to the opposite situation of a time discontinuity (or a time boundary) occurring in a homogeneous and infinite medium. For times $t < 0$, the medium will have a refractive index n_1 everywhere and, for $t = 0$ the refractive index will be suddenly shifted to a new value n_2.

The electric field associated with a plane wave propagating in the medium, with the initial frequency ω_i, will be described by an expression similar to equation (6.1)

$$\vec{E}(x, t) = \begin{cases} \vec{E}_i \exp i(kx - \omega_i t), & (t < 0) \\ \vec{E}_r \exp i(kx - \omega_r t) + \vec{E}_t \exp i(kx - \omega_t t), & (t > 0). \end{cases} \quad (6.10)$$

As before, the three distinct plane waves have to satisfy the linear dispersion relations of the medium, which implies that

$$\omega_i = \frac{kc}{n_1}, \quad \omega_r = -\omega_t, \quad \omega_t = \omega_i \frac{n_1}{n_2}. \quad (6.11)$$

We should take notice of the symmetry between these relations and those of equation (6.2). It can easily be understood that the negative frequency mode corresponds indeed to a reflected wave because, in order to describe a real wave, we have to add to fields (6.1, 6.10) their complex conjugates. This operation will then lead to the appearance of a plane wave with positive frequency and propagating in the opposite direction with respect to that of the initial wave: $\vec{E}_r^* \exp[-i(kx + |\omega_r|t)]$.

In order to determine the relative amplitudes of the transmitted and the reflected waves, we have to establish the time continuity conditions for the electromagnetic field. We can see from Maxwell's equations (6.3, 6.4) that they only contain the time derivatives of the fields $\vec{D}$ and $\vec{B}$. These equations can only stay valid for all times if these two fields are continuously varying in time.

The time continuity conditions are then given by

$$\vec{D}(t = 0^-) = \vec{D}(t = 0^+), \quad \vec{B}(t = 0^-) = \vec{B}(t = 0^+). \quad (6.12)$$

Expressing the fields $\vec{D}$ and $\vec{B}$ in terms of the electric field $\vec{E}$ for the three distinct waves present in equation (6.10), we can rewrite these continuity conditions as

$$n_1^2 E_i = n_2^2 (E_r + E_t), \quad E_i = -E_r + E_t \frac{n_2}{n_1}. \quad (6.13)$$

This means that the well-known continuity conditions (6.7), valid for a sharp spatial boundary in the dielectric properties of the propagating medium, are replaced by these new conditions in the case of a sharp time boundary. The explicit result for the reflected and the transmitted wave amplitudes is

$$R \equiv \frac{E_r}{E_i} = \frac{1}{2}\left(\frac{n_1^2}{n_2^2} - 1\right), \quad T \equiv \frac{E_t}{E_i} = \frac{1}{2}\left(\frac{n_1^2}{n_2^2} + 1\right). \quad (6.14)$$

A simple relation between R and T can also be established here:

$$T + R = \frac{n_1^2}{n_2^2}.$$

(6.15)

These equations can be called the *Fresnel formulae for time reflection*. They are very similar to the usual Fresnel formulae (6.8), but they also show quite surprising differences.

The first surprise is that a time discontinuity can lead to a reflected wave, that is, a wave propagating in the opposite direction with respect to the initial wave. The second is that the amplitudes of both the reflected and the transmitted waves can be very large if the value of the new refractive index n_2 is very low. This can be obtained, for instance, by plasma creation (or flash ionization) with a nearly resonant plasma frequency $\omega_p \sim \omega_i$. This means that we could expect to produce electromagnetic energy out of nearly nothing, in exactly the same way as photons can be created from a vacuum in quantum models. In this case, of course, the amount of energy of the waves created by time reflection would have to be furnished by the external agent responsible for the sudden ionization process.

We should however notice that the limit of $n_2 = 0$ cannot be properly described by the above model because it does not take into account the existence of the electrostatic mode, which can be excited in the medium when its dielectric constant is equal to zero. As we know, for a plasma, this is the electron plasma wave. The field of this electrostatic wave would have to be added to the above continuity conditions. Furthermore, for the plasma case, there are important qualitative differences related, not only to dispersion, but also to the appearance of a magnetic mode, as will be seen later.

A large amount of work has been devoted to the problem of wave transformation due to a sudden creation of a plasma medium, in several different configurations including the presence of a static magnetic field [44]. However, the concept of time refraction was never explored.

6.2   Generalized Fresnel formulae

We can generalize the above treatement of independent space and time discontinuities by considering the case of a sharp boundary of the refractive index moving in space with a constant velocity $\vec{u} = u\vec{e}_x$. An earlier discussion of this problem can be found in reference [110].

Let us assume an incident wave of the form

$$\vec{E}_i \exp i(k_i x - \omega_i t)$$

(6.16)

with $k_i = (\omega_i/c)n_1$. In the reference frame moving with the boundary, the same wave will be represented by

$$\vec{E}_i' \exp i(k_i' x' - \omega_i' t').$$

(6.17)

Let us invoke the Lorentz transformations from the moving frame to the rest frame:

$$x = \gamma(x' + ut'), \quad t = \gamma\left(t' + \frac{\beta}{c}x'\right) \tag{6.18}$$

with $\beta = u/c$ and $\gamma^{-2} = (1 - \beta^2)$.

We know that the field phase in equations (6.16, 6.17) is a relativistic invariant, which means that $k_i x - \omega_i t = k_i' x' - \omega_i' t'$. Replacing equations (6.18) in this equality, and separately equating the terms containing x' and t', we obtain the appropriate Lorentz transformations for the frequency and the wavenumber:

$$\omega' = \gamma\omega_i(1 - \beta n_1), \quad k_i' = \gamma k_i\left(1 - \frac{\beta}{n_1}\right). \tag{6.19}$$

From this we get the dispersion relation in the moving frame

$$k_i' = \frac{\omega'}{c}n_1' \tag{6.20}$$

with

$$n_1' = \frac{n_1 - \beta}{1 - \beta n_1}. \tag{6.21}$$

This equation relates the values of the refractive index as seen in the moving and in the rest frames. It is a very simple and useful result which, surprisingly, cannot easily be found in the literature.

Of course, this transformation is only valid for a non-dispersive medium. The case of a plasma will be considered below. We are now able to repeat, in the moving frame, the above derivation of the Fresnel formulae (6.8). To do this we assume that the total electric field in this frame is formally identical to that of equation (6.1), i.e.

$$\vec{E}'(x', t') = \begin{cases} \vec{E}_i' \exp i(k_i' x' - \omega' t') + \vec{E}_r' \exp i(k_r' x' - \omega' t'), & (x' < 0) \\ \vec{E}_t' \exp i(k_t' x' - \omega' t'), & (x' > 0). \end{cases} \tag{6.22}$$

In analogy with equation (6.2), the three different wavenumbers are related by $k_r' = -k_i'$ and $k_t' = k_i'(n_2'/n_1')$. This leads to the result

$$\frac{E_r'}{E_i'} = \frac{1 - w}{1 + w}, \quad \frac{E_t'}{E_i'} = \frac{2}{1 + w} \tag{6.23}$$

with $w = (k_t'/k_i') = (n_2'/n_1')$.

Let us see how such a result can be extended to the plasma case. To do that we assume that medium 1 is just the vacuum region and medium 2 is a plasma moving in a vacuum with velocity u. This means that the particles contained in the plasma medium (electrons and ions) all move with an average velocity equal to u.

It can easily be realized that the plasma frequency is a relativistic invariant

$$\omega_p^2 = k^2 c^2 - \omega^2 = k'^2 c^2 - \omega'^2. \tag{6.24}$$

This means that, for the plasma case, equation (6.21) is not valid and we have

$$w = \left| \frac{k_t' c}{\omega'} \right| = \sqrt{1 - \frac{\omega_p^2}{\omega'^2}}. \tag{6.25}$$

We clearly see that $w = 0$ defines a cut-off condition and that total reflection will occur for

$$\omega_p^2 = \omega^2 \frac{1 + \beta}{1 - \beta}. \tag{6.26}$$

This is in agreement with our single photon model of chapter 3. The transformations (6.23) stay valid for the moving plasma case if we assume that w is determined by equation (6.25).

We can now make a Lorentz transformation of the electric fields back to the rest frame. For the field of the incident wave, we have

$$E_i' = \gamma(E_i - u B_i) = \gamma E_i (1 - \beta n_1). \tag{6.27}$$

For the reflected wave β is replaced by $-\beta$ and for the transmitted wave n_1 is replaced by n_2. From equations (6.23) we then get

$$R \equiv \frac{E_r}{E_i} = \frac{1 - \beta n_1}{1 + \beta n_1} \frac{1 - w}{1 + w}, \quad T \equiv \frac{E_t}{E_i} = \frac{1 - \beta n_1}{1 - \beta n_1} \frac{2}{1 + w}. \tag{6.28}$$

We can easily find that

$$1 + R = T \frac{1 - \beta n_2}{1 - \beta n_1}. \tag{6.29}$$

These equations can be called the generalized Fresnel formulae for moving dielectric perturbations. We notice that for a stationary boundary, $\beta = 0$, these equations reduce to (6.8, 6.9), as expected.

In the limit of an infinite velocity, $\beta \to \infty$, which would be the case for a time discontinuity, equation (6.29) also reduces to (6.15). However, in this limit, equations (6.28) do not reduce to the Fresnel formulae for time reflection (6.14). This apparent incongruence is due to the fact that the calculations presented in this section are based on Lorentz transformations and therefore they are not valid for $\beta > 1$.

6.3   Magnetic mode

An interesting new effect occurs when the moving discontinuity of the refractive index is associated with an ionization front. We are refering to the excitation of

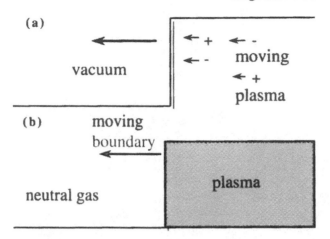

Figure 6.1. Comparison between the two cases: (a) a moving plasma and (b) a moving ionization front.

a purely magnetic mode [53, 97]. This will be examined here in some detail and will be compared with the (at first sight) generic results of the previous section.

Let us study the problem in the frame of the moving ionization front. In this frame, the atoms of the neutral gas region are seen to flow with a velocity $-u$ and to disappear at the front discontinuity $x' = 0$.

On the other hand, the plasma electrons (assuming that they are created with zero kinetic velocity) will appear to flow away from that front with exactly the same speed. The electron equation of motion can be written, in the same reference frame, as

$$\frac{d\vec{v}}{dt'} = -\frac{e}{m\gamma}(\vec{E}' - u\vec{e}_x \times \vec{B}')$$ (6.30)

with

$$\frac{d}{dt'} = \frac{\partial}{\partial t'} - u\frac{\partial}{\partial x'}.$$ (6.31)

Notice that, if instead of an ionization front we were considering a plasma moving with the same velocity, we would have to use $u = 0$ in these equations, because the plasma electrons would be moving with the front. The electric and magnetic fields $\vec{E}'$ and $\vec{B}'$ are determined by Maxwell's equations (6.3), which can now be written as

$$\nabla' \times \vec{E}' = -\mu_0\frac{\partial \vec{H}'}{\partial t'}, \quad \nabla' \times \vec{H}' = \vec{J}' + \epsilon_0\frac{\partial \vec{E}'}{\partial t'}$$ (6.32)

where the electron current is simply

$$\vec{J}' = -en_0\gamma\vec{v}'$$ (6.33)

and n_0 is the electron mean density in the rest frame.

We can now solve these equations by assuming that the fields evolve in space and time according to $\exp i(k'x' - \omega't')$. We obtain

$$i(\omega' + k'u)\vec{v}' = \frac{e}{m\gamma}\left[\vec{E}' - \frac{u}{\omega'}\vec{e}_x \times (\vec{k}' \times \vec{E}')\right] \qquad (6.34)$$

$$i\left[\vec{k}' \times (\vec{k}' \times \vec{E}') + \frac{\omega'^2}{c^2}\right]\vec{E}' = -en_0\gamma\mu_0\omega'\vec{v}'. \qquad (6.35)$$

Notice that we cannot divide the first of these equations by $(\omega' + k'u)$ because such a factor can eventually be equal to zero, as will be shown below. From these two equations we can easily derive the following dispersion relation, valid for transverse modes $(\vec{k}' \cdot \vec{E}' = 0)$ in the plasma region:

$$(k'^2c^2 + \omega_p^2 - \omega'^2)(\omega' + k'u) = 0. \qquad (6.36)$$

This shows that, for a given frequency, two distinct modes are possible:

$$k' = \frac{\omega'}{c}\sqrt{1 - (\omega_p/\omega')^2} \qquad (6.37)$$

and

$$k' = k'_m = -\frac{\omega'}{u}. \qquad (6.38)$$

The first mode is the usual transverse electromagnetic mode in a plasma. The second mode can be called the magnetic mode because its electric field is equal to zero in the rest frame. This can be seen by noting that, according to Maxwell's equations (6.32), we have

$$E'_m = \frac{\omega'}{k'_m}B'_m = -uB'_m. \qquad (6.39)$$

Using the Lorentz transformation for the electric field we get, in the rest frame,

$$E_m = \gamma(E'_m + uB'_m) = 0. \qquad (6.40)$$

Similarly, it can be shown that its frequency in the rest frame is also equal to zero:

$$\omega_m = \gamma(\omega' + uk'_m) = 0, \quad k_m = \gamma\left(k'_m + \beta\frac{\omega'}{c}\right) = \frac{\omega}{c}\frac{1 - \beta}{1 + \beta}. \qquad (6.41)$$

This shows that the magnetic mode is a purely magnetostatic perturbation in the rest frame, with zero frequency but a finite wavelength $2\pi/k_m \neq 0$. From the above discussion we can conclude that the total electric field associated with an ionization front will have four (and not three) distinct waves:

$$\vec{E}'(x',t') = \begin{cases} \vec{E}_i' \exp i(k_i'x' - \omega't') + \vec{E}_r' \exp i(k_r'x' - \omega't'), & (x' < 0) \\ \vec{E}_t' \exp i(k_t'x' - \omega't') + \vec{E}_m' \exp i(k_m'x' - \omega't'), & (x' > 0). \end{cases}$$
(6.42)

The continuity conditions for the transverse fields $\vec{E}'$ and $\vec{B}'$ can then be stated as

$$E_i' + E_r' = E_t' + E_m', \quad E_i' - E_r' = E_t'\frac{k_t'}{k_i'} - E_m'\frac{k_m'}{k_i'}$$
(6.43)

where

$$\frac{k_t'}{k_i'} = w = \sqrt{1 - (\omega_p/\omega')^2}, \quad \frac{k_m'}{k_i'} = -\frac{1}{\beta}.$$
(6.44)

We need to complement the two continuity conditions (6.43) by another condition because we have here three unknown field amplitudes. This extra condition is provided by the second of Maxwell's equations (6.32).

We notice that the current $\vec{J}'$ is continuous across the front boundary if we assume that the electrons are created with zero net velocity: $\vec{J}' = \vec{v}' = 0$ for $x' = 0$. Because the electric field is also continuous, this implies that $\partial B'/\partial x'$ will also be continuous across the boundary:

$$\left(\frac{\partial B'}{\partial x'}\right)_{x'=0^-} = \left(\frac{\partial B'}{\partial x'}\right)_{x'=0^+}.$$
(6.45)

This new continuity condition can be rewritten as

$$k_i'B_i' + k_r'B_r' = k_t'B_t' + k_m'B_m'.$$
(6.46)

Noting that $k_r' = k_i'$, and that $B' = (k'c/\omega')E'$, we can derive from this and from equations (6.43) the three conditions for the electric field amplitudes:

$$E_i' + E_r' = E_t' + E_m'$$

$$E_i' - E_r' = wE_t' + \frac{E_m'}{\beta}$$

$$E_i' + E_r' = w^2 E_t' + \frac{E_m'}{\beta^2}.$$
(6.47)

This can be solved as

$$\frac{E_r'}{E_i'} = \frac{1+\beta}{1-\beta}\frac{1-w}{1+w}$$

$$\frac{E_t'}{E_i'} = \frac{2}{1+\beta w}\frac{1+\beta}{1+w}$$

$$\frac{E_m'}{E_i'} = \frac{2\beta^2(1-w)}{(1-\beta)(1+\beta w)}.$$
(6.48)

This result has to be compared with equations (6.23), which are valid in the absence of the magnetic mode. The difference between these two results can better be understood if we calculate the energy of the reflected wave.

In order to perform this calculation we first notice that the energy of an electromagnetic wave is (apart from the constant $\hbar$, taken here to be equal to 1) the product of the wave frequency with the number of photons: $W = \omega N$. Because the number N is a relativistic invariant, the energy will be Lorentz transformed in a way that is the reverse of the Lorentz transformation for the frequency.

We can then write, for the number of photons associated with both the incident and the reflected wave

$$N_i = \frac{W_i}{\omega} = \frac{W_i'}{\omega'}, \quad N_r = \frac{W_r}{\omega_r} = \frac{W_r'}{\omega'}. \tag{6.49}$$

This means that, for both cases (absence and presence of the magnetic mode), we can write the fraction of reflected energy as

$$\frac{W_r}{W_i} = \frac{\omega_r}{\omega} \frac{W_r'}{W_i'} = \frac{\omega_r}{\omega} \left| \frac{E_r'}{E_i'} \right|^2. \tag{6.50}$$

Using the Lorentz transformation for the the frequencies, we get

$$\omega' = \frac{\omega}{\gamma(1+\beta)} = \frac{\omega_r}{\gamma(1-\beta)} \tag{6.51}$$

which leads to

$$\frac{W_r}{W_i} = \frac{1-\beta}{1+\beta} \left| \frac{E_r'}{E_i'} \right|^2. \tag{6.52}$$

In the absence of the magnetic mode (in the case of a moving plasma, considered in the previous section), and assuming total reflection ($w = 0$), we can obtain from equations (6.23, 6.52)

$$\frac{W_r}{W_i} = \frac{1-\beta}{1+\beta} = \frac{\omega_r}{\omega}. \tag{6.53}$$

This means that, for a counter-propagating ionization front ($\beta < 0$), we have a significant gain in the energy of the reflected wave. From equation (6.49) we can also see that the number of photons is conserved, $N_r = N_i$, which means that all the incident photons are reflected, but with a different frequency.

In contrast, when the magnetic mode is present (in the case of an ionization front, considered in this section), and assuming once again that total reflection is taking place ($w = 0$), we get from equation (6.48)

$$\frac{W_r}{W_i} = \frac{1+\beta}{1-\beta} = \frac{\omega}{\omega_r}. \tag{6.54}$$

This means that, for $\beta < 0$, we now have a significant decrease in the energy of the reflected wave. From equation (6.49) we can also see that the number of photons is not conserved upon reflection, even if it is a total reflection:

$$\frac{N_r}{N_i} = \left(\frac{\omega}{\omega_r}\right)^2.$$
(6.55)

The missing photons are, in this case, converted into a static magnetic field (zero energy photons), and we could be talking about a photon freezing or photon condensation effect. In the next section we will consider a situation which, in some respects, can be considered as the reverse of this photon freezing.

6.4 Dark source

Let us assume a static electric field, applied to a region of space containing a neutral gas. No net current is produced by this field, as long as the atoms or molecules of the gas stay in the neutral state. But, if an ionization front propagates across that region, the static electric field will accelerate the free electrons created by the front, and the resulting space- and time-varying currents will eventually become a source of electromagnetic radiation.

To be specific, we first assume a sinusoidal electrostatic field, described by

$$\vec{E}_0(\vec{r}) = \vec{E}_0 \cos(k_0 x) = \frac{\vec{E}_0}{2}\left(e^{ik_0 x} + \text{c.c.}\right), \quad \vec{B}_0(\vec{r}) = 0.$$
(6.56)

Let us also assume that the ionization front moves with velocity $\vec{u}$ along the x-axis, but in the negative direction, and that the electrostatic field is perpendicular to this velocity:

$$\vec{E}_0 = E_0 \vec{e}_y, \quad \vec{u} = -u\vec{e}_x.$$
(6.57)

Assuming that the ion motion is irrelevant to the fast processes associated with high-frequency wave radiation, we can write the electron current in the plasma region as

$$\vec{J} = -en_0 H(x + ut)\vec{v}$$
(6.58)

where the Heaviside function $H(x + ut)$ represents the sharp boundary between the neutral and the plasma regions, and the electron velocity $\vec{v}$ is determined by the linearized equation of motion

$$\frac{\partial \vec{v}}{\partial t} = -\frac{e}{m}\left[\vec{E}_0(\vec{r}) + \vec{E}\right].$$
(6.59)

Here, $\vec{E}$ represents the radiation field. This field is determined by the wave equation

$$\nabla^2 \vec{E} - \frac{1}{c^2} \frac{\partial^2 \vec{E}}{\partial t^2} = \mu_0 \frac{\partial \vec{J}}{\partial t}. \tag{6.60}$$

Taking the time derivative of equation (6.58) and assuming that the electrons are created at the plasma boundary $x = ut$ with no kinetic energy, we obtain

$$\frac{\partial \vec{J}}{\partial t} = -en_0 u \delta(x + ut)\vec{v} - en_0 H(x + ut)\frac{\partial \vec{v}}{\partial t}$$

$$= \frac{e^2 n_0}{m} H(x + ut)\left[\vec{E}_0(\vec{r}) + \vec{E}\right]. \tag{6.61}$$

If we want to calculate the field radiated in the plasma region $x > ut$ and propagating in the forward direction, along the x-axis and in the positive direction, we can write from these two equations

$$\left(\frac{\partial^2}{\partial x^2} - \frac{1}{c^2}\frac{\partial^2}{\partial t^2} - \frac{\omega_p^2}{c^2}\right)\vec{E} = \frac{\omega_p^2}{c^2} H(x + ut)\vec{E}_0(\vec{r}_\perp, x). \tag{6.62}$$

In order to solve this equation, let us introduce a time Fourier transformation defined by

$$\vec{E}(x, t) = \int \vec{E}_\omega(x) e^{-i\omega t} \frac{d\omega}{2\pi}. \tag{6.63}$$

The one-dimensional wave equation becomes

$$\frac{\partial^2}{\partial x^2}\vec{E}_\omega + \frac{1}{c^2}(\omega^2 - \omega_p^2)\vec{E}_\omega = \frac{\omega_p^2}{c^2}\vec{E}_0(x)\int H(x + ut) e^{i\omega t}\, dt. \tag{6.64}$$

The time integral in the source term can easily be solved and we are led to the following equation:

$$\frac{\partial^2}{\partial x^2}E_\omega + k^2 E_\omega = iA\left\{\exp\left[i\left(k_0 - \frac{\omega}{u}\right)x\right] + \exp\left[-i\left(k_0 + \frac{\omega}{u}\right)x\right]\right\} \tag{6.65}$$

with a wavenumber k determined by

$$k^2 c^2 = \omega^2 - \omega_p^2 \tag{6.66}$$

and a source term amplitude A where

$$A = \frac{1}{2\omega u}\frac{\omega_p^2}{c^2}(\vec{E}_0 \cdot \hat{a}). \tag{6.67}$$

Here we have used the unit polarization vector $\hat{a} = \vec{E}_\omega / |\vec{E}_\omega|$. The solution of equation (6.65), for forward propagation, is

$$E_\omega(x) = \frac{A}{2k} e^{ikx} \int_{-\infty}^{x} e^{-ikx'}\left\{\exp\left[i\left(k_0 - \frac{\omega}{u}\right)x'\right]\right.$$

$$\left. + \exp\left[-i\left(k_0 + \frac{\omega}{u}\right)x'\right]\right\}\, dx'. \tag{6.68}$$

The asymptotic value of the radiated field, observed in the plasma side but very far away from the ionization front ($x \rightarrow \infty$), will contain at most two spectral components:

$$E_\omega(x) = \frac{\pi}{k} A\, e^{ikx} \left[\delta\left(k - k_0 + \frac{\omega}{u} \right) + \delta\left(k + k_0 + \frac{\omega}{u} \right) \right] \qquad (6.69)$$

with the allowed wavenumbers determined by

$$k = -\frac{\omega}{u} \pm k_0. \qquad (6.70)$$

We can see that only the positive sign is allowed here because we are considering forward propagation ($k > 0$). This expression also implies that we have an upper limit for the frequency of the radiation produced in the plasma side: $\omega < k_0 u$.

The value of ω can be obtained by using this equation in the dispersion relation (6.66):

$$\left(k_0 - \frac{\omega}{u} \right)^2 c^2 = \omega^2 - \omega_p^2. \qquad (6.71)$$

The explicit result is

$$\omega = k_0 u \gamma^2 \left[1 - \beta \sqrt{ 1 - \frac{\omega_p^2}{\gamma^2 k_0^2 u^2} } \right] \qquad (6.72)$$

with $\beta = u/c$ and $\gamma^2 = (1 - \beta^2)^{-1}$.

This result was first obtained by Mori *et al* [79], who used a different approach. The interest of our present approach is that we did not use any Lorentz transformation on the front frame. Let us consider the case where $\omega_p^2 \ll \gamma^2 k_0^2 u^2$, which corresponds to a large gamma factor or a small plasma density.

Equation (6.72) then reduces to

$$\omega \simeq \frac{k_0 u}{2} + \frac{\omega_p^2}{2 k_0 u}. \qquad (6.73)$$

This result is very interesting because it shows that a large value of the radiated field frequency ω can be obtained even with very small values of k_0, which means, for a long length scale of the static electric field $\vec{E}_0(x)$, if we use $k_0 u \ll \omega_p$, we obtain a radiation frequency $\omega \simeq \omega_p^2/2 k_0 u \gg \omega_p^2$.

The result described by the present model can be seen as a special case of photon acceleration: (virtual) photons initially with zero frequency, and associated with the static electric field (6.56), are accelerated into high frequencies by the relativistic ionization front and propagate along the plasma medium in the forward direction.

The conversion efficiency of the process is stated by the radiation field solution, equation (6.69):

$$|E_\omega|^2 \sim \left(\frac{\pi}{k}A\right)^2 = \frac{\pi^2}{4k^2}\left(\frac{\omega_p}{c}\right)^4 \frac{|E_0|^2}{\omega^2 u^2}. \tag{6.74}$$

Let us now consider the field radiated into the neutral gas side, for $x < ut$, and propagating in the backward direction. If we restrict again our attention to the one-dimensional problem, this field is determined by the wave equation

$$\left(\frac{\partial^2}{\partial x^2} - \frac{1}{c^2}\frac{\partial^2}{\partial t^2}\right)\vec{E} = \frac{\omega_p^2}{c^2}H(x+ut)\vec{E}_0(x). \tag{6.75}$$

Following the above procedure, we can write the asymptotic solution, valid far away from the ionization front, for $x \to -\infty$, as

$$E_\omega(x) \simeq \frac{\pi}{k}A\,e^{-ikx}\left[\delta\left(k+k_0-\frac{\omega}{u}\right) + \delta\left(k-k_0-\frac{\omega}{u}\right)\right]. \tag{6.76}$$

The two possible spectral components are

$$k = \frac{\omega}{u} \pm k_0. \tag{6.77}$$

In contrast with the previous case where the resulting radiation was propagating in the plasma side, here the waves propagate in a vacuum and have to satisfy the vacuum dispersion relation $\omega = kc$. Using this in the above condition, this leads to

$$\omega = \frac{k_0 u}{1-\beta}. \tag{6.78}$$

This means that, for relativistic ionization fronts such that $(1-\beta) \sim \gamma^2 \gg 1$, very high frequency radiation fields can eventually be emitted. However, the total energy associated with this new wave is much smaller than the energy of the lower frequency emitted into the plasma side, because the value of k appearing in equation (6.76) is now much larger, typically by a factor of γ^2.

The apparatus necessary to produce this kind of radiation process can be based on an array of capacitors, periodically spaced along the x-axis with periodicity $2\pi/k_0$, in order to create the static periodic electric field defined in equation (6.56). Such a configuration can be named a dark source [79] because it transforms a static field into very high frequency photons propagating in both the plasma and the vacuum region.

However, other configurations can also be imagined, based on the same principle, as will be illustrated in the following pages. Let us then generalize the above procedure, by replacing the static periodic electric field by an arbitrary electrostatic field:

$$\vec{E}_0(\vec{r}) = \int \vec{E}_q(\vec{r}_\perp)\,e^{iqx}\,\frac{dq}{2\pi}. \tag{6.79}$$

The previous case obviously corresponds to

$$\vec{E}_q(\vec{r}_\perp) = \pi \vec{E}_0 \left[\delta(q - k_0) + \delta(q + k_0)\right].$$ (6.80)

The one-dimensional wave equation for the time Fourier component of the radiated field, valid in the plasma region for this case of an arbitrary electrostatic field, is

$$\frac{\partial^2}{\partial x^2} E_\omega + \frac{1}{c^2}(\omega^2 - \omega_p^2) E_\omega = i \int A_q \, e^{i(q - \omega/u)x} \frac{dq}{2\pi}$$ (6.81)

with

$$A_q = \frac{\omega}{u} \frac{\omega_p^2}{c^2} \left(\vec{E}_q \cdot \hat{a}\right).$$ (6.82)

The general solution for radiation in the forward direction is then given by

$$E_\omega(x) = \frac{1}{2k} e^{ikx} \int_{-\infty}^{\infty} dx' \int \frac{dq}{2\pi} A_q \, e^{-i(k - q + \omega/u)x'}.$$ (6.83)

Asymptotically, for $x \to \infty$, we get

$$E_\omega(x) = \frac{\pi}{k} A_0 \, e^{ikx}$$ (6.84)

where $A_0 = A_q$ for $q = q_0$, and q_0 is determined by

$$q_0 = k + \frac{\omega}{u}.$$ (6.85)

A similar result could equally well be obtained for radiation emitted in the backward direction and propagating in the vacuum region.

As an example of an application of this more general formulation of the problem, let us consider the case of an electric field acting on a very small region of space, around some point $x = x_0$. This could, for instance, be obtained by a capacitor of infinitesimal width (two pointlike electrodes) located at that position:

$$\vec{E}_0(\vec{r}) = \vec{E}_0 \delta(x - x_0).$$ (6.86)

This can be described by equation (6.79) by using an infinite Fourier spectrum with amplitudes

$$\vec{E}_q(\vec{r}_\perp) = 2\pi \vec{E}_0 \, e^{-iqx_0}.$$ (6.87)

In this case, we could generate an infinitely large spectrum of radiation, in both the forward and the backward direction. Such a configuration, different (but very similar in its principle) from the original idea of a dark source using a capacitor array as described above, could be interesting for the production of ultra-short (or even sub-cycle) radiation pulses, if the resulting broad spectrum of radiation was subsequently compressed by some appropriate optical system.

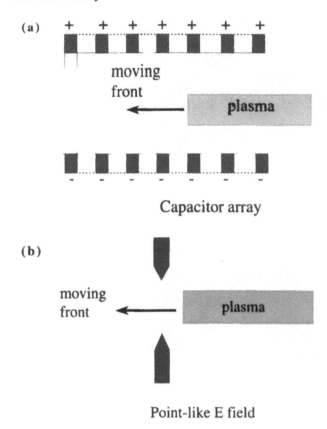

Figure 6.2. Examples of dark sources: (a) a periodic E-field produced by a capacitor array; (b) a pointlike E-field.

Exploring further this idea of producing high-frequency radiation out of a sharp ionization front moving with relativistic velocity, we can even imagine a situation where the static electric field is absent. This means that we do not need to consider any capacitor system acting on the gas medium: $\vec{E}_0(\vec{r}) = 0$. But now, some other ingredient has to be introduced in order to replace this field.

We can, for instance, assume that the density of the background neutral gas medium is modulated in space. Such a modulation can be obtained, for instance, by producing a sound wave, or by forcing the gas to flow through a parallel grid before entering the interaction zone.

When the ionization front moves across the gas, a space modulation of the electron plasma density will take place. The resulting electron current in the plasma region left behind the front is now described by

$$\vec{J} = -en_0[1 + \epsilon \cos(k_0 x)]H(x + ut)\vec{v} \tag{6.88}$$

where ϵ is the amplitude of the density modulation and k_0 determines the period of the space modulation.

A radiation field $\vec{E}$ can eventually result from this configuration, as determined by the wave equation (6.60). But now the electron velocity $\vec{v}$ is determined by equation (6.59) with $\vec{E}_0(\vec{r}) = 0$.

Instead of equation (6.61), we will be led to a Mathieu-type wave equation, which shows unstable solutions inside some region of the space of parameters. In this case, a dark source radiation instability will occur. A brief discussion of the Mathieu equation is postponed to the end of the next chapter.

Chapter 7

Non-stationary processes in a cavity

In this new chapter devoted to the full wave theory of photon acceleration, we will explore a simple theoretical model for mode coupling inside an electromagnetic cavity. This is formally analogous to the elementary quantum theory of collisions and it has the merit of reducing the space–time-varying evolution problem to a purely time-varying problem.

Moreover, this cavity model can also be adequately applied to several experimental configurations. It was initially developed in references [67, 100].

7.1 Linear mode coupling theory

Let us consider an electromagnetic cavity with metalic walls containing a neutral gas, which can be ionized by some external agent (for instance, by an intense laser pulse or by a high voltage applied to adequate electric probes). In quite general conditions, we can describe the electric field associated with the electromagnetic modes contained in the cavity by the following equation:

$$\left(\nabla^2 - \frac{1}{c^2} \frac{\partial^2}{\partial t^2} \right) \vec{E} = \mu_0 \frac{\partial \vec{J}}{\partial t}. \tag{7.1}$$

The electric field has to satisfy certain boundary conditions and its value will be determined by the source term which is described by the electric current $\vec{J}$. The boundary conditions will imply that the general solution of this equation can take the form of a superposition of eigenmodes, such that

$$\vec{E}(\vec{r}, t) = \sum_l e_l(t) \vec{\mathcal{E}}_l(\vec{r}) \tag{7.2}$$

where $e_l(t)$ are time-dependent amplitudes, and $\mathcal{E}_l(\vec{r})$ are the cavity eigenmodes depending on three quantum numbers labelled by l.

In order to be more specific, let us assume the simplest possible example of a rectangular cavity with the length of the three sides given by L_x, L_y and L_z. If we restrict our discussion to transverse electric modes, the fields $\mathcal{E}_l(\vec{r})$ are determined by

$$\mathcal{E}_{lx}(\vec{r}) = \mathcal{E}_0 \cos(m\pi x/L_x) \sin(n\pi y/L_y) \sin(p\pi z/L_z),$$

$$\mathcal{E}_{ly}(\vec{r}) = \mathcal{E}_0 \frac{m}{n} \frac{L_y}{L_x} \sin(m\pi x/L_x) \cos(n\pi y/L_y) \sin(p\pi z/L_z), \qquad (7.3)$$

$$\mathcal{E}_{lz}(\vec{r}) = 0.$$

The quantities m, n and p are integers and, in this case, we have $l \equiv (m, n, p)$. The generalization of the present discussion to another type of cavity, and the inclusion of transverse magnetic modes, is straightforward.

In order to guarantee that the eigenmodes are normed and orthogonal we define the constant $\mathcal{E}_0$ as

$$\mathcal{E}_0 = \sqrt{\frac{8}{V}} \left(1 + \frac{m^2}{n^2} \frac{L_y^2}{L_x^2} \right)^{-1/2} \qquad (7.4)$$

where we have introduced the cavity volume $V = L_x L_y L_z$. This choice for $\mathcal{E}_0$ allows us to write the orthonormality condition

$$\int_V \mathcal{E}_l(\vec{r}) \cdot \mathcal{E}_{l'}(\vec{r}) \, d\vec{r} = \delta_{ll'}. \qquad (7.5)$$

Let us now turn to the electric current appearing in equation (7.1). If we make some simple and plausible assumptions for the gas ionization process, this source term can be written as a function of the cavity electric field $\vec{E}(\vec{r}, t)$ [7].

We first notice that the plasma created out of the neutral gas can be seen as containing an infinity of electronic species, corresponding to the electrons created at different times:

$$\vec{J} = -e \sum_i \Delta n_i \vec{v}_i. \qquad (7.6)$$

The density Δn_i of each of these electron populations is determined by

$$\Delta n_i = \left(\frac{\partial n}{\partial t} \right)_{t_i} \Delta t \qquad (7.7)$$

where Δt is the elementary time interval around t_i during which these populations are created.

The velocity $\vec{v}_i$ of the electrons created at the successive instants t_i is determined by

$$\vec{v}_i(t) = -\frac{e}{m} \int_{t_i}^{t} \vec{E}(t') \, dt' + \vec{v}_{i0}. \qquad (7.8)$$

We can replace this integral inside equation (7.7), assume that the electrons are created initially with zero kinetic energy, $\vec{v}_{i0} = 0$, and take the limit $\Delta t \to 0$. The result is

$$\vec{J} = \frac{e^2}{m} \int_{-\infty}^{t} dt' \left(\frac{\partial n}{\partial t} \right)_{t'} \int_{t'}^{t} dt'' \vec{E}(t''). \qquad (7.9)$$

If we take the time derivative of the current, we get

$$\frac{\partial \vec{J}}{\partial t} = \frac{e^2}{m} \left[\int_{-\infty}^{t} \left(\frac{\partial n}{\partial t} \right)_{t'} \vec{E}(t') \, dt' + \left(\frac{\partial n}{\partial t} \right)_t \int_{t}^{t} \vec{E}(t'') \, dt'' \right]. \qquad (7.10)$$

The second term in this expression is obviously equal to zero, and therefore the equation reduces to

$$\frac{\partial \vec{J}}{\partial t} = \frac{e^2}{m} \int_{-\infty}^{t} \left(\frac{\partial n}{\partial t} \right)_{t'} \vec{E}(t') \, dt'$$

$$= \frac{e^2}{m} \left[n(t') \vec{E}(t') \right]_{-\infty}^{t} - \int_{-\infty}^{t} n(t') \frac{\partial \vec{E}}{\partial t'} \, dt'. \qquad (7.11)$$

If the electric field varies much faster than the electron density (which implies once again the existence of two timescales), this expression reduces to

$$\frac{\partial \vec{J}}{\partial t} = \epsilon_0 \omega_p^2(t) \vec{E}(t). \qquad (7.12)$$

This allows us to write the field equation (7.1) in the following closed form:

$$\left(\nabla^2 - \frac{1}{c^2} \frac{\partial^2}{\partial t^2} \right) \vec{E} = \frac{\omega_p^2}{c^2} \vec{E}. \qquad (7.13)$$

We can now replace the mode decomposition (7.2) in this equation and, after using the orthonormality condition (7.5), we obtain

$$\left(\frac{\partial^2}{\partial t^2} + k_l^2 c^2 \right) e_l(t) = - \sum_{l'} C_{ll'}(t) e_{l'}(t) \qquad (7.14)$$

where we have used

$$k_l^2 \equiv k_{mnp}^2 = (m\pi/L_x)^2 + (n\pi/L_y)^2 + (p\pi/L_z)^2. \qquad (7.15)$$

This equation describes the time evolution of the mode amplitudes due to the linear mode coupling induced by the ionization of the gas inside the cavity. The mode coupling coefficients $C_{ll'}$ are determined by

$$C_{ll'}(t) = \int_V \omega_p^2(\vec{r}, t) \mathcal{E}_l(\vec{r}) \cdot \mathcal{E}_{l'}(\vec{r}) \, d\vec{r}. \qquad (7.16)$$

The mode coupling equation (7.14) can also be written in the following form:

$$\left[\frac{\partial^2}{\partial t^2} + \omega_l^2(t) \right] e_l(t) = - \sum_{l' \neq l} C_{ll'}(t) e_{l'}(t). \qquad (7.17)$$

Here we have used the time-dependent mode frequency

$$\omega_l^2(t) = k_l^2 c^2 + C_{ll}(t). \qquad (7.18)$$

Here, the self-coupling coefficient ($l' = l$) corresponds to a kind of averaged plasma frequency, taken over the mode space configuration

$$C_{ll} = \int_v \omega_p^2(\vec{r}, t)|\mathcal{E}_l(\vec{r})|^2 \, d\vec{r}. \qquad (7.19)$$

The mode coupling equation (7.17) is the basic equation of the present model, and can be used for a detailed description of the spectral changes inside the electromagnetic cavity. Its physical content and its relevance to our problem will be discussed in the next few sections.

7.2   Flash ionization in a cavity

The simplest possible model of plasma formation inside a cavity is to admit that ionization is uniform over the entire volume of the cavity: $\omega_p^2(\vec{r}, t) \equiv \omega_p^2(t)$. In this case, from equation (7.16), we have

$$C_{ll'}(t) = \omega_p^2(t)\delta_{ll'}. \qquad (7.20)$$

This means that the cavity eigenmodes are decoupled from each other and that they evolve according to the equation

$$\frac{\partial^2 e_l}{\partial t^2} + \omega_l^2(t)e_l = 0 \qquad (7.21)$$

where $\omega_l^2(t) = k_l^2 c^2 + \omega_p^2(t)$.

Assuming that the plasma frequency ω_p changes over a slow timescale, we can integrate this equation and obtain the following WKB solution:

$$e_l(t) = E_l\sqrt{k_l c/\omega_l(t)} \exp\left[-i\int^t \omega_l(t') \, dt'\right]. \qquad (7.22)$$

Here E_l is a constant amplitude. In order to be more specific, let us use the following explicit law for the plasma creation inside the cavity:

$$\omega_p^2(t) = \omega_{p0}^2 \left(1 - e^{-\gamma t}\right). \qquad (7.23)$$

This means that the plasma is formed on a timescale of $1/\gamma$. The integral in equation (7.22) can now be written as

$$\int^t \omega_l(t') \, dt' = k_l c \left\{\left[1 + \frac{1}{2}\left(\frac{\omega_{p0}}{k_l c}\right)^2\right]t + \frac{1}{2\gamma}\left(\frac{\omega_{p0}}{k_l c}\right)^2 e^{-\gamma t}\right\}. \qquad (7.24)$$

Let us see what happens during the first stage of the plasma formation when we have $\gamma t \ll 1$. We simply get

$$e_l(t) \simeq E_l \exp\left(-\mathrm{i}k_l ct - \mathrm{i}\frac{\omega_{p0}^2}{2\gamma k_l c}\right). \tag{7.25}$$

This means that, in this short time limit, the cavity eigenmode keeps oscillating at the same frequency, but suffers a small phase shift. In the opposite limit of very long times, such that $\gamma t \gg 1$, we have

$$e_l(t) \simeq E_l\sqrt{k_l c/\omega_l(\infty)}\exp(-\mathrm{i}\omega_l(\infty)t) \tag{7.26}$$

where $\omega_l(\infty)$ is the asymptotic value of the mode eigenfrequency for very large times: $\omega_l^2(\infty) = k_l^2 c^2 + \omega_{p0}^2$.

We can see that this WKB solution gives, for long times, a result which coincides with that of the simple theory of flash ionization outlined in previous chapters: the frequency shift of a given eigenmode of an electromagnetic cavity, where a plasma is uniformly created by some external agent, is simply determined by the asymptotic value of the plasma frequency.

On the other hand, in order to compare this new approach with the previous one, we should not forget that an eigenmode in a cavity is equivalent to two travelling waves propagating in opposite directions. This can be seen from equations (7.3), where the factor $\sin(p\pi z/L_z)$ can be decomposed into the two factors $\exp(\pm\mathrm{i}k_z z)$, with $k_z = p\pi/L_z$, representing propagation along the z-axis in opposite directions.

However, as we will see next, this view of flash ionization is still too simple and we have to refine it. A more realistic model for flash ionization inside a cavity will have to contain the description of some spatial structure. In general, the plasma creation will not be completely uniform over the entire cavity volume, and we have to retain the space dependence of $\omega_p^2(\vec{r}, t)$. This is illustrated in figure 7.1.

Let us then return to the mode coupling equation (7.17) and try a WKB solution of the form (7.22), but where E_l represents a time-dependent amplitude. Taking the second time derivative of this new solution we get

$$\frac{\partial^2 e_l}{\partial t^2} \simeq -\omega_l^2 e_l - 2\mathrm{i}\sqrt{\omega_l k_l c}\frac{\partial E_l}{\partial t}\exp\left[-\mathrm{i}\int^t \omega_l(t')\,\mathrm{d}t'\right]. \tag{7.27}$$

Using this in the mode coupling equation, we obtain an equation for the slowly varying amplitude, in the form

$$\frac{\partial E_l}{\partial t} = -\mathrm{i}\sum_{l\neq l'} B_{ll'}(t)E_{l'}. \tag{7.28}$$

The new mode coupling coefficients are determined by

$$B_{ll'}(t) = \frac{C_{ll'}(t)}{2\sqrt{\omega_l(t)\omega_{l'}(t)}}\exp\left(-\mathrm{i}\int^t [\omega_{l'}(t') - \omega_l(t')]\,\mathrm{d}t'\right). \tag{7.29}$$

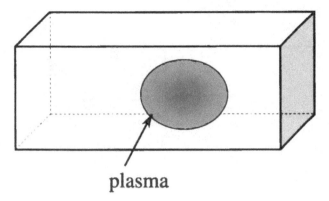

plasma

Figure 7.1. Non-uniform plasma formation inside a cavity.

Let us examine the simple but physically meaningful case where a given mode $l = l_0$ pre-exists in the cavity for $t < 0$, and all the other modes are absent before the plasma formation. If the coupling is not too strong, the amplitude of this mode can still be considered as approximately constant for times $t \geq 0$.

Using this parametric approximation, we can write from equation (7.27) that, for any other mode $l \neq l_0$, the amplitudes are

$$E_l(t) = -iE_{l_0} \int^t B_{ll'}(t')\,dt'. \tag{7.30}$$

This equation allows us to calculate explicitly, and in a simple way, the mode coupling efficiency. The explicit calculation can be performed by using the following model for non-uniform ionization inside the cavity:

$$\omega_p^2(\vec{r}, t) = \omega_{p0}^2(1 - e^{-\gamma t})\exp[-a(\vec{r} - \vec{r}_0)^2]. \tag{7.31}$$

Here $1/\gamma$ is the timescale for plasma formation and $a^{-1/2}$ is the dimension of the ionized region around some point $\vec{r}_0$. According to their definition, the mode coupling coefficients $C_{ll'}(t)$ can be split into two factors, one containing the time dependence and the other containing the spatial mode coupling.

Using equations (7.31) and (7.16), we obtain

$$C_{ll'}(t) = \omega_{p0}^2(1 - e^{-\gamma t})I_{ll'} \tag{7.32}$$

where

$$I_{ll'} = \int^t e^{-a(\vec{r} - \vec{r}_0)^2}\mathcal{E}_l(\vec{r}) \cdot \mathcal{E}_{l'}(\vec{r})\,d\vec{r}. \tag{7.33}$$

Using the eigenmodes (7.3), or any others, this quantity can then be explicitly evaluated. From our analysis, two new aspects relevant to flash ionization can be clearly identified [67].

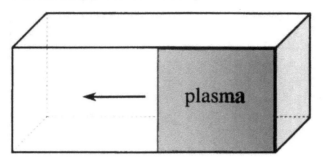

Figure 7.2. Moving ionization front inside a cavity.

First, the ionization processes inside a cavity not only produce a shift in the frequency of a pre-existing mode but also give rise to a large spectrum of radiation associated with the amplitudes $e_l(t)$, for $l \neq l_0$, due to linear mode coupling. Second, the frequency shift associated with each mode is not simply determined by the asymptotic value of the plasma frequency ω_{p0} but by the self-coupling coefficient $C_{ll} < \omega_{p0}$ defined by equation (7.19), which is a kind of plasma frequency averaged over the eigenmode spatial field distribution.

This means that the spectrum generated by flash ionization inside a cavity is broader, and the frequency shift of the pre-excited mode is smaller, than predicted by single photon dynamics. These two qualitative aspects are well confirmed by experimental results in microwave cavities, where the observed frequency shifts were always smaller than the expected $\omega_l = \sqrt{k_l^2 c^2 + \omega_{p0}^2}$, and a broad spectrum of radiation was observed.

7.3 Ionization front in a cavity

Let us now turn to the problem of an ionization front travelling across the cavity, as illustrated in figure 7.2. Then $\omega_p^2(\vec{r}, t)$ will be of the general form $\omega_p^2(\vec{r} - \vec{v}t)$. In the simplest but physically realistic case of a front which is uniform in the planes $z = $ const, and starts its motion at the position $z = L_z$, moving along the z-axis with a velocity v, we can write

$$\omega_p^2(\vec{r}, t) = \omega_{p0}^2 f(z + vt) \tag{7.34}$$

where $f(z + vt)$ describes the form of the front.

In this case, the coupling coefficients (7.16) can be written as

$$C_{ll'}(t) = \frac{2}{L_z} \omega_{p0}^2 \delta_{mm'} \delta_{nn'} I_{pp'}(t) \tag{7.35}$$

where the integral $I_{pp'}(t)$ is determined by

$$I_{pp'}(t) = \int_0^{L_z} \sin(p\pi z/L_z) \sin(p'\pi z/L_z) f(z + vt) \, dz. \tag{7.36}$$

In order to have an order of magnitude estimate of the amplitude of the cavity modes excited by the ionization front, we can use the simple model for the front form

$$f(z + vt) = H(z + vt - L_z) \tag{7.37}$$

where $H(z - a)$ is the Heaviside function and the particular value $a = L_z - vt$ is chosen in order to describe a sharp front starting its motion at $(z = L_z, t = 0)$ and moving across the cavity along the z-axis towards the point $z = 0$ with a negative velocity.

For this particular case, the integral (7.36) can be easily calculated and gives

$$\begin{aligned}
I_{pp'}(t) = &- \frac{L_z}{2\pi} \frac{1}{p - p'} \sin\left[(p - p')\pi(1 - vt/L_z)\right] \\
&+ \frac{L_z}{2\pi} \frac{1}{p + p'} \sin\left[(p + p')\pi(1 - vt/L_z)\right].
\end{aligned} \tag{7.38}$$

It can be noticed from equation (7.35) that the mode coupling is only associated with the quantum numbers p and p', leaving the other two quantum numbers m and n invariant. This is due to the fact that the plasma frequency is constant along the x- and the y-axis with which these two quantum numbers are associated. This means that the mode coupling will only occur along the direction of the electron density gradient.

Using this expression in equation (7.35) and noting that $k_z = p\pi/L_z$ and $k_z' = p'\pi/L_z$, we obtain

$$\begin{aligned}
C_{ll'}(t) = &-\frac{\omega_{p0}^2}{\pi} \delta_{mm'}\delta_{nn'} \\
&\times \left\{ \frac{(-1)^{p-p'}}{p - p'} \sin[(k_z - k_z')vt] - \frac{(-1)^{p+p'}}{p + p'} \sin[(k_z + k_z')vt] \right\}.
\end{aligned} \tag{7.39}$$

This expression is valid for $p \neq p'$. In the case of $p = p'$, it is replaced by

$$C_{ll}(t) = -\frac{\omega_{p0}^2}{2p\pi} \left\{ 2k_z + \sin[2k_z L_z(1 - vt/L_z)] \right\}. \tag{7.40}$$

In order to discuss the physical meaning of this result, let us go back to the evolution equation (7.17). Let us also assume that the mode l', pre-existing in the cavity for times $t < 0$, is still the dominant mode for subsequent times (which implies that the mode coupling is not very efficient).

For our qualitative discussion it is appropriate to neglect the slow variation of the frequency eigenmode, such that we can use the approximate expression

$$e_{l'}(t) \simeq E' e^{-i\omega_{l'}t}. \tag{7.41}$$

We can then see that only four distinct forced terms exist, for which the mode amplitudes e_l are resonantly excited, and these terms verify the resonance condition

$$\omega_l = \omega_{l'} \pm [(k_z \pm k_z')v]t. \tag{7.42}$$

If we were in a free space configuration, and not inside a cavity, we would simply have $k_z = \omega_l/c$ and $k_z' = \omega_{l'}/c$, which means that this resonance condition would reduce to the well-known expression for the relativistic mirror:

$$\omega_l = \omega_{l'} \frac{1 \pm \beta}{1 \pm \beta} \tag{7.43}$$

where $\beta = v/c$.

The only and significant difference with respect to the relativistic mirror effect is that here we generally have partial and not total reflection. The four resonance conditions therefore have a very simple explanation: the pre-existing cavity mode would correspond in free space to two travelling waves with the same frequency, but propagating in opposite directions. Each of these travelling waves interacts with the front and, from its partial reflection, two new waves result, propagating in opposite directions and with distinct frequencies.

Let us come back to the mode coupling equation and retain one of the four resonant terms. We can then write

$$\left(\frac{d^2}{dt^2} + \omega_l^2 \right) e_l = E' e^{-i\omega_l t} \tag{7.44}$$

where E' depends on the coupling with the pre-existing mode l', and ω_l is given by one of the possible four choices of equation (7.42).

Using

$$e_l(t) = E_l(t) e^{-i\omega_l t} \tag{7.45}$$

we obtain the following saturation amplitude for the resonant mode $p \gg p'$, after the completed journey of the ionization front across the cavity:

$$E_l(\text{sat}) \simeq \frac{\omega_{p0}^2}{4\pi} \frac{E'}{p\omega_l} \frac{L_z}{v}. \tag{7.46}$$

We can see that a larger value for the front velocity v corresponds to a larger value for the quantum number p compatible with the resonance condition, and to a smaller saturation amplitude. This qualitative discussion suggests that the resonant mode with a lower frequency shift will have a larger saturation amplitude than the resonant mode with a larger frequency shift. This eventually explains the absence of this mode in the first experiments of ionization fronts in a microwave cavity [95].

But, apart from the resonant modes which are more strongly coupled with the pre-existing mode in the cavity, many other modes can also be excited by

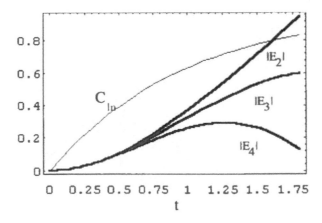

Figure 7.3. Time variation of the mode coupling coefficient and of amplitudes of the lowest order excited modes.

linear mode coupling, leading to an electromagnetic energy cascade over a large spectral width, as shown by numerical integration of the above mode coupling equations [100], and qualitatively confirmed by experiments. See figure 7.3 for a numerical example.

7.4 Electron beam in a cavity

Instead of considering ionization processes inside a cavity, we can also obtain a similar moving boundary if an electron beam is sent across the cavity. The mode coupling due to an electron beam is formally analogous to that due to an ionization front. However, the two processes also present very distinctive features.

First of all, it should be noticed that the field equation (7.1) can also be written as

$$\left(c^2 \nabla^2 - \frac{\partial^2}{\partial t^2}\right) \vec{A} = -\vec{J} \tag{7.47}$$

where $\vec{A}$ is the vector potential.

In our discussion of the flash ionization and the ionization front we used equation (7.12), which can be rewritten as

$$\vec{J} = -\frac{e^2}{m} \int_{-\infty}^{t} \left(\frac{\partial n}{\partial t}\right)_{t'} [\vec{A}(t) - \vec{A}(t')] \, dt'. \tag{7.48}$$

Using this expression in equation (7.47) leads to our previous field equation (7.13). However, the current associated with an electron beam cannot be described by the same source current because equations (7.12, 7.48) were derived for electrons created at specific times and positions, with a velocity equal to zero.

In contrast, the electrons of an electron beam will exist for all times, and will always move with a non-negligible velocity. Using the equation of motion for the beam electrons we can easily obtain the following new expression for the electric current:

$$\vec{J} = -en(\vec{r}, t)\vec{v} = -\frac{e^2}{m}n(\vec{r}, t)\vec{A}. \tag{7.49}$$

Using it in equation (7.47), we obtain

$$\left(c^2\nabla^2 - \frac{\partial^2}{\partial t^2}\right)\vec{A} = \omega_p^2(\vec{r}, t)\vec{A}. \tag{7.50}$$

This is formally identical to equation (7.13), but with the electric field replaced by the vector potential. Because the spatial eigenfunctions for the electric field and for the vector potential are identical, we can again use a linear expansion for $\vec{A}$ of the form

$$\vec{A}(\vec{r}, t) = \sum_l a_l(t)\mathcal{E}_l(\vec{r}). \tag{7.51}$$

The evolution equation for the mode amplitudes can now readily be found:

$$\left(\frac{\partial^2}{\partial t^2} + \omega_l^2(t)\right)a_l(t) = -\sum_{l' \neq l} C_{ll'}(t)a_{l'}(t). \tag{7.52}$$

We see that it is formally identical to equation (7.17), but with a different physical meaning, the mode amplitudes $a_l(t)$ being related to the vector potential eigenmodes. We can then repeat the calculations of the previous section. The result is that the frequency spectum generated by the electron beam will be identical to that generated by the ionization front, because the modes with the same frequency are coupled in the same way.

However, the energy content of each mode is different because the vector potential amplitudes are related to the electric field amplitude by $e_l(t) = a_l(t)\omega_l$. This means that, for instance, the saturation amplitude obtained after the completed journey of the electron beam front across the cavity is now determined by

$$A_l(\text{sat}) \simeq \frac{\omega_{p0}^2}{4\pi} \frac{A'}{p\omega_l} \frac{L_z}{v} \tag{7.53}$$

where A' is the amplitude of the source term and where we have assumed that $a_l(t) \simeq A_l \exp(-i\omega_l t)$.

Again, this is formally identical to equation (7.46), with the electric field amplitudes replaced by vector potential amplitudes. But, if we rewrite this expression in terms of the electric field amplitudes, we realize that the saturation amplitude is now larger, by a factor of $\omega_{l'}/\omega_l$, than that for the ionization front case. Also, as we have seen from the resonance condition for the frequency shift (7.42, 7.43), this factor can be much larger than one.

This means that the electron beam is a much more efficient way to up-shift the electromagnetic energy in the frequency domain, not because the frequency up-shift itself is different, but because the amplitudes of the shifted modes are much larger. This can be illustrated by a numerical integration of the coupled mode equations for both cases [100]. Using the same argument we can also conclude that the down-shifted modes will have, in contrast, a much lower amplitude in the case of the electron beam.

The physical differences between these two processes of frequency up-shifting can be better understood if we consider the total magnetic field generated inside the cavity. We know that the magnetic field is determined by $\vec{B} = \nabla \times \vec{A}$ and, using equation (7.51), we can write, for the case of the electron beam

$$\vec{B}(\vec{r}, t > L_z/v) = \sum_l [\nabla \times \mathcal{E}_l(\vec{r})] A_l \exp(-i\omega_l t). \tag{7.54}$$

This means that the total magnetic field is just the sum of the magnetic fields associated with each of the modes excited in the cavity.

A different situation occurs for the case of an ionization front. If we write the total magnetic field in terms of the total electric field, we get

$$\vec{B}(\vec{r}, t) = -\nabla \times \int_0^t \vec{E}(\vec{r}, t) = -\sum_l [\nabla \times \mathcal{E}_l(\vec{r})] \int_0^t e_l(t')\, dt'. \tag{7.55}$$

The asymptotic value for this magnetic field will then be

$$\vec{B}(\vec{r}, t > L_z/v) = \sum_l \frac{1}{i\omega_l} [\nabla \times \mathcal{E}_l(\vec{r})]$$
$$\times \left(\delta_{ll'} + i\omega_l \int_0^{L_z/v} e_l(t')\, dt' - E_l\, e^{-i\phi_l} + E_l\, e^{-i\omega_l t} \right) \tag{7.56}$$

where we have $\phi_l = \omega_l L_z/v$.

It can be seen from this expression that the last term, which oscillates at the frequencies ω_l, corresponds to the sum of the magnetic fields associated with each of the modes excited in the cavity. But, the remaining terms describe a static magnetic field which was excited by the ionization process and which can be identified with the magnetic mode discussed in the previous chapter.

This means that, in the case of the ionization front, a large amount of energy is transferred to the static magnetic field, which explains why this process of frequency up-shifting is much less efficient than that of an electron beam where such a static field is absent.

Finally, we should keep in mind that other possible non-stationary processes occurring inside the cavity, which do not create new electron populations, are described by the same mode coupling equations as the electron beam case. This is, for instance, the situation where, in a pre-existing plasma medium, a space–time variation of the plasma density is produced, associated with an electron plasma wave with relativistic velocity (wakefield). As shown in the above discussion, the up-shifting frequency effect will be as efficient as in the case of the electron beam and no static magnetic field will be excited.

7.5   Fermi acceleration in a cavity

The linear mode coupling formalism explored in this chapter allows us to give a new description of the Fermi acceleration of photons, using a full wave description [101]. Let us then assume that, due to some externally applied perturbation, the neutral gas inside the cavity is ionized in a limited domain near the boundary $z = L_z$, in such a way that the width of the plasma region oscillates in time with a frequency Ω, much smaller than the cavity mode eigenfrequencies ω_l.

This can be described by the following space–time distribution for the electron plasma frequency:

$$\omega_p^2(\vec{r}, t) = \omega_{p0}^2 H[z - L_z + A(1 - \cos \Omega t)]. \tag{7.57}$$

Here we have used the Heaviside function $H(z - a)$. This simple law describes a plasma with constant plasma frequency ω_{p0} but oscillating in a region of the cavity between L_z and $L_z - 2A$.

The coupling coefficients (7.39) now take the form, for $p \neq p'$,

$$C_{ll'} = -\frac{\omega_{p0}^2}{\pi} \delta_{mm'} \delta_{nn'} \left\{ \frac{1}{p - p'} \sin[\pi(p - p')(1 - \epsilon(1 - \cos \Omega t))] \right.$$
$$\left. - \frac{1}{p + p'} \sin[\pi(p + p')(1 - \epsilon(1 - \cos \Omega t))] \right\} \tag{7.58}$$

where $\epsilon = A/L_z$. For $p = p'$, equation (7.41) becomes

$$C_{ll} = \epsilon \omega_{p0}^2 (1 - \cos \Omega t) - \frac{\omega_{p0}^2}{2p\pi} \sin[2p\pi\epsilon(1 - \cos \Omega t)]. \tag{7.59}$$

Using this explicit expression in the mode coupling equations (7.17) we can calculate the time evolution of the eigenmode field amplitudes. Let us start by neglecting the mode coupling and let us study the evolution of a single mode $e_l(t)$, when according to equation (7.59), the effective value of plasma frequency is modulated by a much lower frequency Ω. Notice that mode coupling would be absent if the plasma frequency were oscillating uniformly over the entire cavity volume.

In this case, we can write the single mode evolution equation as

$$\frac{d^2}{dt^2} e_l(t) + \left\{ k_l^2 c^2 + \epsilon \omega_{p0}^2 (1 - \cos \Omega t) \right.$$

$$+ \frac{\epsilon}{2p\pi} \sin[2p\pi\epsilon(1 - \cos \Omega t)]\bigg\} e_l(t) = 0. \tag{7.60}$$

Here it is appropriate to use a dimensionless time variable $\tau = \Omega t/2$, and to introduce two parameters:

$$\delta = \frac{4}{\Omega^2}(k_l^2 + \epsilon\omega_{p0}^2), \quad \gamma_0 = 2\epsilon\frac{\omega_{p0}^2}{\Omega^2}. \tag{7.61}$$

Equation (7.60) can then be reduced to a more familiar form:

$$\frac{d^2}{d\tau^2}e_l(\tau) + \bigg\{\delta - 2\gamma_0 \cos(2\tau) - 2\frac{\gamma_0}{2p\pi\epsilon} \sin[2p\pi\epsilon(1 - \cos(2\tau))]\bigg\} e_l(\tau) = 0. \tag{7.62}$$

For high-frequency cavity modes such that $2p\pi\epsilon \gg 1$ the last term can be neglected and this reduces to the well-known Mathieu equation. In order to illustrate the main physical processes in the cavity let us first restrict our discussion to this equation.

It is well known [81] that the Mathieu equation has bounded solutions, as well as unbounded ones, depending on the values of the two parameters δ and γ_0. The existence of unbounded (or unstable) solutions is physically very interesting because this means that it is possible to excite cavity modes, starting from the noise level up to arbitrary amplitudes. The energy source responsible for the creation of this electromagnetic field, nearly out of nothing, can only be identified with the external source producing the plasma oscillations.

The stability diagram of the Mathieu equation shows that the most favourable modes to be excited verify the resonance condition $k_l c \simeq m\Omega/2$, where m is some positive integer. However, because we are assuming that Ω is much smaller than the eigenmode frequencies ω_l, this resonance condition implies that m has to be quite large. Also, in this case, we known from the properties of the Mathieu equation that the unstable domain in the parameter space (δ, γ_0) is extremely narrow and very difficult to experimentally satisfy. This means that, in principle, generation of electromagnetic energy inside an oscillating and empty cavity is possible, but it is very unlikely to be observed, due to the fact that it will only occur inside very narrow regions of the parameter space.

We now turn to the stable regions of the parameter space, where it is possible to derive approximate analytical expressions for the time evolution of the amplitude of a given uncoupled mode $e_l(t)$, and to describe the entire frequency spectrum associated with this single mode. Due to the frequency modulations of the self-coupling coefficient (7.59), the frequency spectrum of a single mode no longer corresponds to a well-defined frequency ω_l, but to an infinite number of frequencies.

This can be clearly seen if we assume that the mode amplitude $e_l(t)$ is described by

$$e_l(t) = \sum_k e_{lk} \exp[-i(\bar{\omega}_l + k\Omega)t] \tag{7.63}$$

where we have used $\bar{\omega}_l^2 = k_l^2 c^2 + \epsilon \omega_{p0}^2$, and have assumed that the amplitudes $e_{lk}(t)$ vary over a timescale much larger than $1/\bar{\omega}_l$.

Using this solution in the single mode evolution equation (7.62), and retaining only the terms with the same fast timescale, we can easily derive an equation for the new amplitudes $e_{lk}(t)$:

$$
\begin{aligned}
& - 2i(\bar{\omega}_l + k\Omega)\frac{d}{dt}e_{lk} + [\bar{\omega}_l^2 - (\bar{\omega}_l + k\Omega)^2]e_{lk} \\
& - \frac{\epsilon}{2i}\omega_{p0}^2(e_{l,k-1} - e_{l,k+1}) + i\frac{\omega_{p0}^2}{4p\pi}\sum_m J_m(2p\pi\epsilon) \\
& \times \left[i^m e^{2ip\pi\epsilon}e_{l,k-m} - i^{-m} e^{-2ip\pi\epsilon}e_{l,k+m}\right] = 0.
\end{aligned}
\tag{7.64}
$$

In order to obtain this expression we have used the expansion of the sine term of equation (7.62) in Bessel functions of integer order J_m. We can simplify it considerably if we notice that, for $\Omega \ll \bar{\omega}_l$, the second term (proportional to e_{lk}) can be neglected. On the other hand, if the oscillation parameter ϵ is small and if the quantum number p is large, in such a way that the inequality $2p\pi\epsilon \ll 1$ is satisfied, we can also neglect the term containing the sum over the Bessel functions.

With these two simplifying assumptions, we can reduce equation (7.64) to

$$
\frac{d}{dt}e_{lk} = \epsilon\frac{\omega_{p0}^2}{4\omega_k}(e_{l,k-1} - e_{l,k+1})
\tag{7.65}
$$

where we have used $\omega_k = \bar{\omega}_l + k\Omega$.

This system of coupled equations for the amplitudes e_{lk} can be easily integrated by noting that it is formally identical to the well-known recurrence relation for Bessel functions J_k [84]. This means that it satisfies the following solution:

$$
e_{lk}(t) = J_k\left(\epsilon\frac{\omega_{p0}^2}{2\omega_k}t\right).
\tag{7.66}
$$

Let us now consider the opposite limit of small ϵ and large p, such that $2p\pi\epsilon \gg 1$. In this case, the sum term dominates over the term proportional to ϵ in equation (7.64). Moreover, we also have $J_1(2p\pi\epsilon) \gg J_m(2p\pi\epsilon)$, for $m \neq 1$, which means that we can also neglect all the terms in the series expansion except those for $m = \pm 1$.

The resulting equation is

$$
\frac{d}{dt}e_{lk} = i\frac{\omega_{p0}^2}{4p\pi\omega_k}J_1(2p\pi\epsilon)\cos(2p\pi\epsilon)(e_{l,k-1} + e_{l,k+1}).
\tag{7.67}
$$

Here again, the integration becomes possible because of the formal analogy with the recurrence relation between the Bessel functions with purely imaginary argument I_k. The solution can now be written as

$$e_{lk}(t) = (-1)^k \mathrm{i}^{-k} J_k \left(\frac{\omega_{p0}^2}{2p\pi \omega_k} \cos(2p\pi \epsilon) J_1(2p\pi \epsilon)t \right). \qquad (7.68)$$

We clearly see from these solutions that, in the two opposite limits of the parameter $(2p\pi \epsilon)$, an infinitely large spectrum of frequencies with decreasing amplitudes can be generated out of a single mode contained in a variable cavity. This means that Fermi acceleration results in a considerable line broadening.

This effect of energy spreading over the frequency domain is even more dramatic when we include the linear mode coupling associated with the coefficients $C_{ll'}$ established above. Once again, it is not possible to integrate the resulting mode equations, unless we introduce some simplifying assumptions.

First of all, let us assume that the cavity contains a dominant mode characterized by the index l', such that coupling occurs mainly between this and the other cavity modes. We can write the amplitude $e_{l'}$ of the dominant mode in the form

$$e_{l'}(t) = E_{l'} \, \mathrm{e}^{-\mathrm{i}\bar{\omega}_{l'}t} \qquad (7.69)$$

where $E_{l'}$ can be assumed nearly constant.

The amplitudes of all the other modes, $l \neq l'$, can then be described by

$$\left[\frac{\mathrm{d}^2}{\mathrm{d}t^2} + \omega_l^2(t) \right] e_l(t) = -C_{ll'} E_{l'} \, \mathrm{e}^{-\mathrm{i}\bar{\omega}_{l'}t} \qquad (7.70)$$

where $\omega_l^2(t)$ and $C_{ll'}(t)$ are determined by equations (7.18, 7.58).

We can see from here that the modes which are excited with a larger amplitude satisfy the nearly resonant condition

$$\omega_l(t) \simeq \bar{\omega}_l = \bar{\omega}_{l'} \pm m\Omega \qquad (7.71)$$

where m is an integer.

For these nearly resonant modes, we can write the following evolution equation:

$$\left[\frac{\mathrm{d}^2}{\mathrm{d}t^2} + \omega_l^2(t) \right] e_l(t) = \pm \mathrm{i} E_{l'} \frac{\omega_{p0}^2}{2\pi} \left\{ \frac{J_m((p+p')\pi \epsilon)}{p+p'} \right.$$
$$\left. - \frac{J_m((p-p')\pi \epsilon)}{p-p'} \right\} \mathrm{e}^{-\mathrm{i}\bar{\omega}_{l'}t}. \qquad (7.72)$$

This equation can easily be integrated if we neglect the slow time evolution of the eigenfrequency $\omega_l \simeq \bar{\omega}_l$, and assume that

$$e_l(t) = E_l \, \mathrm{e}^{-\mathrm{i}\bar{\omega}_l t} \qquad (7.73)$$

where $E_l(t)$ is the slow amplitude of the nearly resonant modes.

In this approximation, the solution for short times is given by

$$E_l(t) = \mp E_{l'} \frac{\omega_{p0}^2}{4\pi \bar{\omega}_l} \left\{ \frac{J_m((p+p')\pi \epsilon)}{p+p'} - \frac{J_m((p-p')\pi \epsilon)}{p-p'} \right\} t. \qquad (7.74)$$

This result shows that the modes that are more easily excited are those for which, in addition to the resonance condition (7.71), we have $p \simeq p'$. Of course, for low modulation frequencies $\Omega \ll \omega_l$, these resonance conditions are satisfied by several modes. In the limit $\Omega \ll (\omega_{l+1} - \omega_l)$ all the modes in the cavity will be nearly resonant, with an amplitude approximately given by equation (7.74).

If we complement this with the tendency of each mode considered individually to suffer a considerable resonant broadening, as shown by the above discussion of the single mode evolution, we conclude that Fermi acceleration inside an oscillating cavity provides an efficient mechanism for generating a broad spectrum of electromagnetic radiation.

This is in qualitative agreement with our previous discussion of the Fermi photon acceleration using the Hamiltonian approach, where it was shown that the photon trajectories would become stochastic under certain conditions. Here the particle stochastic behaviour is replaced by the mode coupling, in the same way as stochasticity in classical particle motion is replaced by a non-stochastic description based on the wave equation for a quantum particle.

Chapter 8

Quantum theory of photon acceleration

In order to complete our theoretical framework, it is important to show that photon acceleration is not a result of the classical or semi-classical approximations, but that it can easily be identified and described in purely quantum grounds.

As an introduction to this chapter, we will describe first the well-known quantization procedure of the electromagnetic field [37, 61], with small changes, in order to adapt the notation to our specific needs. We will also describe the less well-known procedure for the field quantization in a plasma.

We will then examine the quantum theory of space and time refraction, which not only confirms the main features of the classical theory, but also identifies intrinsic quantum processes such as the possibility of photon creation from the vacuum.

8.1 Quantization of the electromagnetic field

8.1.1 Quantization in a dielectric medium

We consider an infinite, non-dispersive and non-dissipative dielectric medium, characterized by a real dielectric constant, ϵ. We give a phenomenological description of the problem of the field quantization in this medium, by assuming that ϵ is a well-known constant, given *a priori* and not explicitly calculated by a microscopic theory. This approach will be useful for our problem.

We start with Maxwell's equations which can be written, in the absence of charge and current distributions, as

$$\nabla \times \vec{E} = -\frac{\partial \vec{B}}{\partial t}, \quad \nabla \cdot \vec{D} = 0$$

$$\nabla \times \vec{H} = \frac{\partial \vec{D}}{\partial t}, \quad \nabla \cdot \vec{B} = 0$$

$$(8.1)$$

with $\vec{D} = \epsilon \vec{E}$ and $\vec{B} = \mu_0 \vec{H}$.

We define the scalar and the vector potentials, ϕ and $\vec{A}$, such that

$$\vec{E} = -\frac{\partial \vec{A}}{\partial t} - \nabla \phi, \quad \vec{B} = \nabla \times \vec{A}. \tag{8.2}$$

It is convenient to choose the Coulomb gauge, defined by the condition

$$\nabla \cdot \vec{A} = 0. \tag{8.3}$$

In this case, we can derive from equations (8.1) the following equations for the scalar and vector potentials:

$$\nabla^2 \phi = 0 \tag{8.4}$$

$$\nabla^2 \vec{A} - \frac{n^2}{c^2} \frac{\partial^2 \vec{A}}{\partial t^2} = \frac{n^2}{c^2} \nabla \frac{\partial \phi}{\partial t} \tag{8.5}$$

with $n = \sqrt{\epsilon/\epsilon_0}$.

This obviously implies that $\phi = 0$, and

$$\nabla^2 \vec{A} - \frac{n^2}{c^2} \frac{\partial^2 \vec{A}}{\partial t^2} = 0. \tag{8.6}$$

We know that, in classical theory, the general solution of this equation can be written as

$$\vec{A}(\vec{r}, t) = 2 \int \vec{A}_k \, e^{i(\vec{k} \cdot \vec{r} - \omega_k t)} \frac{d\vec{k}}{(2\pi)^3} \tag{8.7}$$

where $\omega_k = (kc/n)$.

In this expansion, the integration extends over both the negative and the positive values of the three components of the wavevector $\vec{k}$. But we know that, in order to guarantee that the electromagnetic fields are real quantities, we have to assume that $\vec{A}_{-k} = \vec{A}_k^*$.

This allows us to rewrite the above general solution as

$$\vec{A}(\vec{r}, t) = \int \vec{A}_k(\vec{r}, t) \frac{d\vec{k}}{(2\pi)^3} \tag{8.8}$$

with

$$\vec{A}_k(\vec{r}, t) = \left[\vec{A}_k \, e^{i(\vec{k} \cdot \vec{r} - \omega_k t)} + \vec{A}_k^* \, e^{-i(\vec{k} \cdot \vec{r} - \omega_k t)} \right]. \tag{8.9}$$

The corresponding electric and magnetic fields are, according to equations (8.2),

$$\vec{E}(\vec{r}, t) = \int \vec{E}_k(\vec{r}, t) \frac{d\vec{k}}{(2\pi)^3}, \quad \vec{B}(\vec{r}, t) = \int \vec{B}_k(\vec{r}, t) \frac{d\vec{k}}{(2\pi)^3} \tag{8.10}$$

with

$$\vec{E}_k(\vec{r}, t) = -\mathrm{i}\omega_k \left[\vec{A}_k \, \mathrm{e}^{\mathrm{i}(\vec{k}\cdot\vec{r}-\omega_k t)} - \vec{A}_k^* \, \mathrm{e}^{-\mathrm{i}(\vec{k}\cdot\vec{r}-\omega_k t)} \right] \qquad (8.11)$$

$$\vec{B}_k(\vec{r}, t) = \mathrm{i}\vec{k} \times \left[\vec{A}_k \, \mathrm{e}^{\mathrm{i}(\vec{k}\cdot\vec{r}-\omega_k t)} - \vec{A}_k^* \, \mathrm{e}^{-\mathrm{i}(\vec{k}\cdot\vec{r}-\omega_k t)} \right]. \qquad (8.12)$$

Let us now consider the total electromagnetic energy density

$$W = \int W_k \frac{\mathrm{d}\vec{k}}{(2\pi)^3} \qquad (8.13)$$

with

$$W_k = \frac{1}{2} \left[\epsilon |\vec{E}_k(\vec{r}, t)|^2 + \frac{1}{\mu_0} |\vec{B}_k(\vec{r}, t)|^2 \right] = 2\epsilon\omega_k^2(\vec{A}_k \cdot \vec{A}_k^*). \qquad (8.14)$$

We know that the amplitudes of the vector potential are, in general, complex quantities and it is useful to consider their real and imaginary parts separately, by defining

$$q_k = \sqrt{\epsilon}(A_k + A_k^*), \quad p_k = \mathrm{i}\omega_k\sqrt{\epsilon}(A_k - A_k^*). \qquad (8.15)$$

Here we have considered scalar quantities, by using $\vec{A}_k = A_k\vec{e}_k$, where $\vec{e}_k$ is the unit polarization vector. From these two quantities, we can write

$$A_k = \frac{1}{4\epsilon\omega_k}(\omega_k q_k + \mathrm{i}p_k). \qquad (8.16)$$

Using this in equation (8.14), we obtain for the energy density of the mode $\vec{k}$

$$W_k = \frac{1}{2}(p_k^2 + \omega_k^2 q_k^2). \qquad (8.17)$$

This is formally identical to the energy of a one-dimensional oscillator with unit mass, frequency ω, position q and momentum p:

$$W \equiv H(q, p) = \frac{p^2}{2} + \omega^2\frac{q^2}{2}. \qquad (8.18)$$

The first term represents the kinetic energy and the second term the parabolic potential. This classical oscillator can be quantized by defining a Hamiltonian operator $\hat{H}$, such that

$$\hat{H} = \frac{1}{2}(\hat{p}^2 + \omega^2\hat{q}^2) \qquad (8.19)$$

where the position and the momentum operators satisfy the commutation relations

$$[\hat{q}, \hat{p}] \equiv \hat{q}\hat{p} - \hat{p}\hat{q} = \mathrm{i}\hbar. \qquad (8.20)$$

The Hamiltonian operator can also be written in terms of the destruction and creation operators, $\hat{a}$ and $\hat{a}^+$, such that

$$\hat{q} = \sqrt{\frac{\hbar}{2\omega}}(\hat{a} + \hat{a}^+), \quad \hat{p} = -i\sqrt{\frac{\hbar\omega}{2}}(\hat{a} - \hat{a}^+). \qquad (8.21)$$

This is equivalent to

$$\hat{a} = \frac{1}{\sqrt{2\hbar\omega}}(\omega\hat{q} + i\hat{p}), \quad \hat{a}^+ = \frac{1}{\sqrt{2\hbar\omega}}(\omega\hat{q} - i\hat{p}). \qquad (8.22)$$

From this and from equations (8.17, 8.20), we have

$$\hat{a}^+\hat{a} = \frac{1}{2\hbar\omega}(\omega^2\hat{q}^2 + \hat{p}^2 + i\omega[\hat{q}, \hat{p}]) = \frac{1}{\hbar\omega}\left(\hat{H} - \frac{1}{2}\hbar\omega\right). \qquad (8.23)$$

We also have

$$\hat{a}\hat{a}^+ = \frac{1}{\hbar\omega}\left(\hat{H} + \frac{1}{2}\hbar\omega\right). \qquad (8.24)$$

This leads to the following commutation relation:

$$[\hat{a}, \hat{a}^+] = \hat{a}\hat{a}^+ - \hat{a}^+\hat{a} = 1. \qquad (8.25)$$

From equations (8.19, 8.24) we can also write the Hamiltonian operator in the form

$$\hat{H} = \hbar\omega\left(\frac{1}{2} + \hat{N}\right) \qquad (8.26)$$

where $\hat{N}$ is the quantum number operator

$$\hat{N} = \hat{a}^+\hat{a}. \qquad (8.27)$$

Returning to the electromagnetic field, we see that we can associate with each mode $\vec{k}$ a one-dimensional quantum oscillator with destruction and creation operators $\hat{a}_k$ and $\hat{a}_k^+$, such that the field amplitude operators for each mode are established by the equivalence

$$\vec{A}_k \to \sqrt{\frac{\hbar}{2\epsilon\omega_k}}\hat{a}_k\vec{e}_k, \quad \vec{A}_k^* \to \sqrt{\frac{\hbar}{2\epsilon\omega_k}}\hat{a}_k^+\vec{e}_k^*. \qquad (8.28)$$

At this point we should notice that the electromagnetic radiation has two independent polarization vectors. In order to take them both into account we change the operators in the following way:

$$\hat{a}_k\vec{e}_k \to \sum_{\lambda=1,2} a(\vec{k}, \lambda)\vec{e}(\vec{k}, \lambda). \qquad (8.29)$$

From now on, we will remove the hat in the operator notation. The two unit polarization vectors are orthogonal:

$$\vec{e}(\vec{k}, \lambda) \cdot \vec{e}(\vec{k}, \lambda') = \delta_{\lambda\lambda'}. \tag{8.30}$$

An obvious choice of these two orthogonal polarization vectors is two linear polarizations in the plane perpendicular to the direction of wave propagation:

$$\vec{k} \cdot \vec{e}(\vec{k}, \lambda) = 0 \quad (\lambda = 1, 2) \tag{8.31}$$

$$\vec{e}(\vec{k}, 1) \times \vec{e}(\vec{k}, 2) = \vec{k}/|k|. \tag{8.32}$$

The total vector potential operator will then be given by

$$\vec{A}(\vec{r}, t) = \sum_{\lambda=1,2} \int \frac{d\vec{k}}{(2\pi)^3} \sqrt{\frac{\hbar}{2\epsilon\omega_k}} \left[a(\vec{k}, \lambda)\vec{e}(\vec{k}, \lambda) e^{i(\vec{k}\cdot\vec{r}-\omega_k t)} \right.$$
$$\left. + a^+(\vec{k}, \lambda)\vec{e}^*(\vec{k}, \lambda) e^{-i(\vec{k}\cdot\vec{r}-\omega_k t)} \right]. \tag{8.33}$$

From equations (8.2) we can also establish the electric and the magnetic field operators:

$$\vec{E}(\vec{r}, t) = i \sum_{\lambda=1,2} \int \frac{d\vec{k}}{(2\pi)^3} \sqrt{\frac{\hbar\omega_k}{2\epsilon}} \left[a(\vec{k}, \lambda)\vec{e}(\vec{k}, \lambda) e^{i(\vec{k}\cdot\vec{r}-\omega_k t)} \right.$$
$$\left. - a^+(\vec{k}, \lambda)\vec{e}^*(\vec{k}, \lambda) e^{-i(\vec{k}\cdot\vec{r}-\omega_k t)} \right] \tag{8.34}$$

and

$$\vec{B}(\vec{r}, t) = i \sum_{\lambda=1,2} \int \frac{d\vec{k}}{(2\pi)^3} \sqrt{\frac{\hbar}{2\epsilon\omega_k}} \left[a(\vec{k}, \lambda)\vec{k} \times \vec{e}(\vec{k}, \lambda) e^{i(\vec{k}\cdot\vec{r}-\omega_k t)} \right.$$
$$\left. - a^+(\vec{k}, \lambda)\vec{k} \times \vec{e}^*(\vec{k}, \lambda) e^{-i(\vec{k}\cdot\vec{r}-\omega_k t)} \right]. \tag{8.35}$$

These field operators are strictly equivalent to those obtained for propagation in a vacuum, the only difference being the replacement of the vacuum permitivity ϵ_0 by the appropriate dielectric constant of the medium, ϵ.

8.1.2 Quantization in a plasma

We consider an infinite, homogeneous and unmagnetized plasma. In contrast with the previous case, we are now dealing with a dispersive medium. Instead of using the plasma dielectric function, it is more appropriate to describe this medium as a vacuum plus charge and current distributions. Maxwell's equations, which are the starting point of our quantization procedure, can then be written in the following form:

$$\nabla \times \vec{E} = -\frac{\partial \vec{B}}{\partial t}, \quad \nabla \cdot \vec{E} = \frac{\rho}{\epsilon_0} \qquad (8.36)$$

$$\nabla \times \vec{H} = \vec{J} + \epsilon_0 \frac{\partial \vec{E}}{\partial t}, \quad \nabla \cdot \vec{B} = 0 \qquad (8.37)$$

with $\vec{B} = \mu_0 \vec{H}$.

Assuming that the plasma ions are at rest, with a mean density n_0, we can write the charge and current densities as

$$\vec{J} = -en\vec{v}, \quad \rho = -e(n - n_0). \qquad (8.38)$$

The electron density and mean velocity are described by the continuity and the momentum conservation equations:

$$\frac{\partial n}{\partial t} + \nabla \cdot n\vec{v} = 0 \qquad (8.39)$$

$$\frac{\partial \vec{v}}{\partial t} + \vec{v} \cdot \nabla \vec{v} = -\frac{e}{m}(\vec{E} + \vec{v} \times \vec{B}) - S_e^2 \nabla \ln n. \qquad (8.40)$$

We define here the electron thermal velocity by the quantity $v_e = \sqrt{T/m}$, where T is the plasma temperature, and use $S_e^2 = 3v_e^2$. By treating the plasma electrons as a fluid we are neglecting the resonant particle effects which can, for instance, lead to purely kinetic effects such as the electron Landau damping. The inclusion of these effects would require a more refined and purely microscopic quantization procedure.

Apart from the fluid approach, we will also restrict our discussion to low field intensities. This allows us to linearize the fluid equations (8.39, 8.40) around the equilibrium state.

Using $n = n_0 + \tilde{n}$, where $|\tilde{n}| \ll n_0$ is the density perturbation, we get

$$\frac{\partial \tilde{n}}{\partial t} + n_0 \nabla \cdot \vec{v} = 0$$

$$\frac{\partial \vec{v}}{\partial t} = -\frac{e}{m}\vec{E} - \frac{S_e^2}{n_0}\nabla n. \qquad (8.41)$$

The linearized electron current and charge densities are

$$\vec{J} = -en_0\vec{v}, \quad \rho = -e\tilde{n}. \qquad (8.42)$$

Let us introduce the scalar and vector potentials ϕ and $\vec{A}$, by using equations (8.2), and retain the Coulomb gauge, defined by the condition (8.3). The corresponding potential equations can be derived from equations (8.37) and take the form

$$\nabla^2 \phi = \frac{e}{\epsilon_0} \tilde{n} \tag{8.43}$$

$$\left(\nabla^2 - \frac{1}{c^2} \frac{\partial^2}{\partial t^2}\right) \vec{A} = \frac{en_0}{c^2 \epsilon_0} \vec{v}_\perp \tag{8.44}$$

where $\vec{v}_\perp$ is the transverse part of the electron velocity, determined by the condition

$$\nabla \cdot \vec{v}_\perp = 0. \tag{8.45}$$

On the other hand, from the electron fluid equations we can easily derive

$$\left(\frac{\partial^2}{\partial t^2} - S_e^2 \nabla^2\right) \tilde{n} = -\frac{en_0}{m} \nabla^2 \phi \tag{8.46}$$

$$\frac{\partial^2 \vec{v}_\perp}{\partial t^2} = \frac{e}{m} \frac{\partial^2 \vec{A}}{\partial t^2}. \tag{8.47}$$

This allows us to write

$$\left(\frac{\partial^2}{\partial t^2} - S_e^2 \nabla^2\right) \tilde{n} = -\omega_p^2 \tilde{n} \tag{8.48}$$

$$\vec{v}_\perp = \frac{e}{m} \vec{A} \tag{8.49}$$

which, according to equations (8.43, 8.44), implies that

$$\left(\nabla^2 - \frac{1}{S_e^2} \frac{\partial^2}{\partial t^2}\right) \phi = \frac{\omega_p^2}{S_e^2} \phi \tag{8.50}$$

$$\left(\nabla^2 - \frac{1}{c^2} \frac{\partial^2}{\partial t^2}\right) \vec{A} = \frac{\omega_p^2}{c^2} \vec{A}. \tag{8.51}$$

We notice here that, if the electron thermal velocity could be made equal to $c/\sqrt{3}$, such that $S_e^2 = c^2$, these two equations would reduce to a single Klein–Gordon equation for the four-vector potential $A^\nu \equiv (\phi/c, \vec{A})$, in the form

$$\left(\nabla^2 - \frac{1}{c^2} \frac{\partial^2}{\partial t^2}\right) A^\nu = \frac{m_{\text{eff}} c}{\hbar} A^\nu \tag{8.52}$$

where $m_{\text{eff}} = \omega_p \hbar / c^2$ would be the mass of the vector field.

However, it is obvious that we have $S_e^2 \neq c^2$ for general plasma conditions, which means that the electromagnetic field in a plasma will not be exactly equivalent (but it will be similar) to a massive vector field. In order to prepare for the field quantization procedure we can write the general solution of equations (8.50, 8.51) in the form

$$\vec{A}(\vec{r}, t) = \int \vec{A}_k(\vec{r}, t) \frac{d\vec{k}}{(2\pi)^3}, \quad \phi(\vec{r}, t) = \int \phi(\vec{r}, t) \frac{d\vec{k}}{(2\pi)^3} \tag{8.53}$$

with

$$\vec{A}_k(\vec{r}, t) = \left[\vec{A}_k \, e^{i(\vec{k}\cdot\vec{r} - \omega_k t)} + \vec{A}_k^* \, e^{-i(\vec{k}\cdot\vec{r} - \omega_k t)} \right] \tag{8.54}$$

and

$$\phi_k(\vec{r}, t) = \left[\phi_k \, e^{i(\vec{k}\cdot\vec{r} - \omega_k t)} + \phi_k^* \, e^{-i(\vec{k}\cdot\vec{r} - \omega_k t)} \right]. \tag{8.55}$$

These two equations for the vector and the scalar potentials are formally identical to each other. But it should be noticed that, according to equations (8.50, 8.51), the frequencies for the vector potential (the first equation) are determined by the dispersion relation

$$\omega_k = \sqrt{k^2 c^2 + \omega_p^2} \tag{8.56}$$

and for the scalar potential (the second equation) by

$$\omega_k = \sqrt{k^2 S_e^2 + \omega_p^2}. \tag{8.57}$$

The corresponding electric and magnetic fields are still represented by equations (8.2, 8.10), but now equation (8.12) is replaced by

$$\vec{E}_k(\vec{r}, t) = i\omega_k \vec{A}_k(\vec{r}, t) - i\vec{k}\phi_k(\vec{r}, t), \quad \vec{B}_k(\vec{r}, t) = i\vec{k} \times \vec{A}_k(\vec{r}, t). \tag{8.58}$$

Let us now consider the total energy density

$$w_k = \frac{1}{2} \left[\epsilon_0 |\vec{E}_k(\vec{r}, t)|^2 + \frac{1}{\mu_0} |\vec{B}_k(\vec{r}, t)|^2 \right] + w_{\text{part}}(\vec{k})$$
$$= 2\epsilon k^2 \phi_k \phi_k^* + 2\omega_k^2 \vec{A}_k \cdot \vec{A}_k^*. \tag{8.59}$$

Here, the first term corresponds to longitudinal (or electrostatic) oscillations, and the second term to the transverse electromagnetic waves. For convenience, we have added to the purely electromagetic energy the kinetic energy of the particles associated with the oscillations of the scalar potential.

It is well known from plasma theory that, in the electron plasma oscillations, the averaged energy is equally divided between the electrostatic field energy and the kinetic energy of the plasma electrons. The result is the appearance of a factor of 2 in the first term of this equation.

Field quantization is obtained by introducing destruction and creation operators $a(\vec{k}, \lambda)$ and $a^+(\vec{k}, \lambda)$, for each of the three distinct modes, such that we have, for the transverse modes ($\lambda = 1, 2$),

$$\vec{A}_k \rightarrow \sqrt{\frac{\hbar}{2\epsilon_0 \omega_k}} a(\vec{k}, \lambda) \vec{e}(\vec{k}, \lambda), \quad \vec{A}_k^* \rightarrow \sqrt{\frac{\hbar}{2\epsilon_0 \omega_k}} a^+(\vec{k}, \lambda) \vec{e}^*(\vec{k}, \lambda) \tag{8.60}$$

and, for the longitudinal mode ($\lambda = 3$),

$$\phi_k \rightarrow \sqrt{\frac{\hbar}{2\epsilon_0 k}} a(\vec{k}, \lambda), \quad \phi_k^* \rightarrow \sqrt{\frac{\hbar}{2\epsilon_0 k}} a^+(\vec{k}, \lambda). \tag{8.61}$$

The total vector potential operator will then be determined by

$$\vec{A}(\vec{r}, t) = \sum_{\lambda=1,2} \int \frac{d\vec{k}}{(2\pi)^3} \sqrt{\frac{\hbar}{2\epsilon_0 \omega_k}} \left[a(\vec{k}, \lambda)\vec{e}(\vec{k}, \lambda) e^{i(\vec{k}\cdot\vec{r}-\omega_k t)} \right.$$
$$\left. + a^+(\vec{k}, \lambda)\vec{e}^*(\vec{k}, \lambda) e^{-i(\vec{k}\cdot\vec{r}-\omega_k t)} \right]. \tag{8.62}$$

The total scalar potential operator is

$$\phi(\vec{r}, t) = \int \frac{d\vec{k}}{(2\pi)^3} \sqrt{\frac{\hbar}{2\epsilon_0 k}} \left[a(\vec{k}, 3) e^{i(\vec{k}\cdot\vec{r}-\omega_k t)} + a^+(\vec{k}, 3) e^{-i(\vec{k}\cdot\vec{r}-\omega_k t)} \right]. \tag{8.63}$$

The resulting total energy operator will be given by

$$W = \sum_{\lambda=1}^{3} \int \frac{d\vec{k}}{(2\pi)^3} \hbar\omega_k(\lambda) \left[a^+(\vec{k}, \lambda)a(\vec{k}, \lambda) + \frac{1}{2} \right] \tag{8.64}$$

such that

$$\omega_k(\lambda) = \sqrt{k^2 c^2(\lambda) + \omega_p^2} \tag{8.65}$$

with $c^2(\lambda = 1, 2) = c^2$, and $c^2(\lambda = 3) = S_e^2$.

The first two modes correspond to the two transverse photons and the third mode corresponds to plasmons (or longitudinal photons). Here we find the three independent polarization states of a massive vector field (which has spin one) [88], but with two distinct characteristic velocities: for the transverse particles it is the speed of light c, and for the plasmons it is the thermal velocity S_e.

The similarities of the electromagnetic field quantization in a plasma with the quantization of a massive vector field was noticed long ago by Anderson [5]. It has served as a phenomenological model for the theory of the Higgs boson [88].

8.2   Time refraction

8.2.1   Operator transformations

Let us consider a time discontinuity in an infinite dielectric medium such that, at time $t = 0$, the dielectric constant suddenly changes from a value ϵ_1 to a new value ϵ_2. This transformation law can be described by the expression

$$\epsilon(t) = \epsilon_1 H(-t) + \epsilon_2 H(t) \tag{8.66}$$

where $H(t)$ is the Heaviside function.

We know from equation (8.34) that, for a given polarization state of the photons in the dielectric medium ($\lambda = 1$, or $\lambda = 2$), the electric field operator can be written as

$$\vec{E}(\vec{r}, t) = i\sqrt{\frac{\hbar \omega_k}{2\epsilon}} \left[a(\vec{k}, t) e^{i\vec{k}\cdot\vec{r}} - a^+(\vec{k}, t) e^{-i\vec{k}\cdot\vec{r}} \right] \vec{e}_k. \tag{8.67}$$

Here we have used a real polarization vector $\vec{e}_k = \vec{e}_k^*$, and introduced time-dependent destruction and creation operators

$$a(\vec{k}, t) = a(\vec{k}) e^{-i\omega_k t}, \quad a^+(\vec{k}, t) = a^+(\vec{k}) e^{i\omega_k t}. \tag{8.68}$$

We also know that the displacement vector and the magnetic field operators can be determined by

$$\vec{D}(\vec{k}, t) = \epsilon \vec{E}(\vec{k}, t), \quad \vec{B}(\vec{k}, t) = \frac{\vec{k}}{\omega_k} \times \vec{E}(\vec{k}, t). \tag{8.69}$$

This means that we can write

$$\vec{D}(\vec{k}, t) = i\vec{e}_k \sqrt{\frac{\hbar}{2}\omega_j \epsilon_j} \left[a_j(\vec{k}, t) e^{i\vec{k}\cdot\vec{r}} - a_j^+(\vec{k}, t) e^{-i\vec{k}\cdot\vec{r}} \right] \tag{8.70}$$

$$\vec{B}(\vec{k}, t) = i(\vec{k} \times \vec{e}_k) \sqrt{\frac{\hbar}{2\omega_j \epsilon_j}} \left[a_j(\vec{k}, t) e^{i\vec{k}\cdot\vec{r}} - a_j^+(\vec{k}, t) e^{-i\vec{k}\cdot\vec{r}} \right]. \tag{8.71}$$

For $t < 0$ we use the index $j = 1$, and for $t > 0$, we use $j = 2$, in these expressions. Our main problem is to relate the new operators a_2 and a_2^+, to the old ones, a_1 and a_1^+. These operators are different from each other because the meaning of a photon (or of an elementary excitation of the field) changes with the refractive index at $t = 0$.

In order to obtain such a relation we use the continuity conditions for the fields

$$\vec{D}(\vec{r}, t = 0^-) = \vec{D}(\vec{r}, t = 0^+), \quad \vec{B}(\vec{r}, t = 0^-) = \vec{B}(\vec{r}, t = 0^+). \tag{8.72}$$

Noting that these equalities are independent of $\vec{r}$, we can easily reduce them to the following relations between the new and the old operators:

$$\sqrt{\omega_1 \epsilon_1}[a_1(\vec{k}) - a_1^+(-\vec{k})] = \sqrt{\omega_2 \epsilon_2}[a_2(\vec{k}) - a_2^+(-\vec{k})] \tag{8.73}$$

$$\frac{1}{\sqrt{\omega_1 \epsilon_1}}[a_1(\vec{k}) + a_1^+(-\vec{k})] = \frac{1}{\sqrt{\omega_2 \epsilon_2}}[a_2(\vec{k}) + a_2^+(-\vec{k})]. \tag{8.74}$$

Let us define the parameter

$$\alpha = \sqrt{\frac{\omega_1 \epsilon_1}{\omega_2 \epsilon_2}} = \sqrt{\frac{n_2 \epsilon_1}{n_1 \epsilon_2}} = \sqrt{\frac{n_1}{n_2}}. \tag{8.75}$$

Equations (8.74) can be rewritten as

$$\alpha[a_1(\vec{k}) - a_1^+(-\vec{k})] = a_2(\vec{k}) - a_2^+(-\vec{k})$$
$$a_1(\vec{k}) + a_1^+(-\vec{k}) = \alpha[a_2(\vec{k}) + a_2^+(-\vec{k})]. \tag{8.76}$$

By adding and subtracting these two equations, we obtain

$$a_1(\vec{k}) = Aa_2(\vec{k}) - Ba_2^+(-\vec{k})$$
$$a_1^+(-\vec{k}) = Aa_2^+(-\vec{k}) - Ba_2(\vec{k}). \tag{8.77}$$

Here we have used the real coefficients A and B, defined by

$$A = \frac{1 + \alpha^2}{2\alpha}, \quad B = \frac{1 - \alpha^2}{2\alpha}. \tag{8.78}$$

The reciprocal relations can also be obtained:

$$a_2(\vec{k}) = Aa_1(\vec{k}) + Ba_1^+(-\vec{k})$$
$$a_2^+(-\vec{k}) = Aa_1^+(-\vec{k}) + Ba_1(\vec{k}). \tag{8.79}$$

This shows that each field mode existing for $t < 0$, with a given wavevector $\vec{k}$, will be coupled to two modes existing for $t > 0$, with wavevectors $\vec{k}$ and $-\vec{k}$. Such a coupling provides an explanation, at the quantum level, of the effect of time reflection obtained with the classical theory of chapter 6.

8.2.2 Symmetric Fock states

It is particularly important to look at the behaviour of the number states, or Fock states, when we go through a time discontinuity. They are defined as the eigenstates of the number operator $N_k \equiv a^+(\vec{k})a(\vec{k})$:

$$a^+(\vec{k})a(\vec{k})|n_k\rangle = n_k|n_k\rangle. \tag{8.80}$$

The eigenvalues n_k of the number operator are the occupation numbers, or the number of photons in the mode $\vec{k}$. The energy eigenvalues are

$$\langle n_k|H_k|n_k\rangle = \hbar\omega_k \left\langle n_k\left| \left(a^+(\vec{k})a(\vec{k}) + \frac{1}{2} \right) \right| n_k \right\rangle = \hbar\omega_k \left(n_k + \frac{1}{2} \right). \tag{8.81}$$

The ground state, corresponding to an occupation number equal to zero, $n_k = 0$, is called the vacuum state. It is well known (and can easily be derived from the above equations) that the eigenvectors $|n_k\rangle$, corresponding to an arbitrary excited state, can be derived by applying n_k times the creation operator $a^+(\vec{k})$ to the vacuum eigenstate $|n_k = 0\rangle \equiv |0_k\rangle$.

For normalized eigenvectors we can write

$$|n_k\rangle = \frac{1}{\sqrt{n!}}[a^+(\vec{k})]^{n_k}|0_k\rangle. \tag{8.82}$$

We have shown that, in the process of time discontinuity, the modes $\vec{k}$ and $-\vec{k}$ are coupled to each other. It is then useful to introduce the following symmetric state vectors:

$$|n, n'\rangle_j \equiv |n_k, n'_{-k}\rangle_j = |n_k\rangle_j |n'_{-k}\rangle_j. \qquad (8.83)$$

The index $j = 1$ pertains to the Fock states valid for $t < 0$, and $j = 2$ to the Fock states valid for $t > 0$. Using equation (8.71), we can then define these symmetric Fock states in terms of the symmetric vacuum:

$$|n, n'\rangle_j = \frac{1}{\sqrt{n!n'!}} [a_j^+(\vec{k})]^n [a_j^+(-\vec{k})]^{n'} |0, 0\rangle_j. \qquad (8.84)$$

This expression means that, if we want to establish a relation between some initial symmetric state $|n, n'\rangle_1$ and some final symmetric state $|m, m'\rangle_2$, we have to establish a relation between the symmetric vacuum states, before and after the time discontinuity, $|0, 0\rangle_1$ and $|0, 0\rangle_2$. This can be achieved by representing the initial vacuum states $|0, 0\rangle_1$ in terms of the final state vectors

$$|0, 0\rangle_1 = \sum_{m,m'} C_{m,m'} |m, m'\rangle_2. \qquad (8.85)$$

It is obvious that such a development has to be perfectly symmetric with respect to the modes $\vec{k}$ and $-\vec{k}$, simply because the vacuum is a state of zero total momentum. This implies that

$$C_{m,m'} = C_m \delta_{mm'}. \qquad (8.86)$$

If we apply the destruction operator $a_1(\vec{k})$ to equation (8.85), and use equation (8.77), we obtain

$$a_1(\vec{k})|0, 0\rangle_1 \equiv 0 = \sum_m C_m \left[A a_2(\vec{k}) - B a_2^+(-\vec{k}) \right] |m, m\rangle_2$$

$$= C_m \left[A\sqrt{m}|m - 1, m\rangle_2 - B\sqrt{m+1} \right] |m, m + 1\rangle_2$$

$$= (A C_{m+1} - B C_m)\sqrt{m+1}|m, m + 1\rangle_2. \qquad (8.87)$$

This leads to the following recurrence relation for the coefficients C_m:

$$C_m = \frac{B}{A} C_{m-1} = \left(\frac{B}{A} \right)^m C_0. \qquad (8.88)$$

The initial symmetric vacuum states (8.85) can then be represented as

$$|0, 0\rangle_1 = C_0 \sum_{m=0}^{\infty} \left(\frac{B}{A} \right)^m |m, m\rangle_2. \qquad (8.89)$$

Now, we can use the normalization condition in order to determine the remaining coefficient:

$$_1\langle 0,0|0,0\rangle_1 = |C_0|^2 \sum_{m=0}^{\infty} \left(\frac{B}{A}\right)^{2m} = |C_0|^2 \frac{1}{1-(B/A)^2} = 1. \tag{8.90}$$

This means that

$$C_0 = e^{i\theta_0}\sqrt{1-(B/A)^2} \tag{8.91}$$

where θ_0 is an arbitrary phase.

Using equation (8.89), and assuming that $\theta_0 = 0$, we can write the final expression for the symmetric vacuum decomposition (8.85) as

$$|0,0\rangle_1 = \sqrt{1-(B/A)^2} \sum_{m=0}^{\infty} \left(\frac{B}{A}\right)^m |m,m\rangle_2. \tag{8.92}$$

This result shows that a time boundary at $t = 0$ will be able to generate, from an initial vacuum state described by the vector states $|0,0\rangle_1$, a number of $2m$ photons which will appear in symmetric pairs of modes, $\vec{k}$ and $-\vec{k}$. According to the above equation, the probability of finding such a symmetric Fock state with $2m$ photons at times $t > 0$, will be given by

$$p(m) =_2 \langle m,m|0,0\rangle_{11}\langle 0,0|m,m\rangle_2 = \left[1-\left(\frac{B}{A}\right)^2\right]\left(\frac{B}{A}\right)^{2m}. \tag{8.93}$$

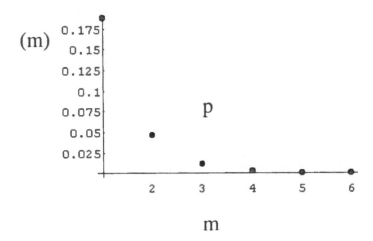

Figure 8.1. Probability of photon creation from the vacuum, for $B/A = 0.5$.

In order to have a more precise idea of the physical consequences of this result, let us assume the following plausible situation of a very small time perturbation of the refractive index of the medium, such that $n_2 = n_1(1 + \delta)$, with $\delta \ll 1$:

$$\alpha = \sqrt{\frac{n_1}{n_2}} \simeq 1 - \frac{\delta}{2}. \tag{8.94}$$

It means that we have $(B/A) \simeq (\delta/2)$, which leads to

$$p(m) \simeq \left(\frac{\delta}{2}\right)^{2m}. \tag{8.95}$$

This gives an order of magnitude estimate for the probability of creation of photons out of the vacuum due to a sudden change of the refractive index. This probability decreases with the number of photons m according to a power law. The possible generation of photons out of the vacuum inside an optical cavity with a time-dependent dielectric was recently considered in reference [19].

8.2.3   Probability for time reflection

The above results on the transformation of an initial vacuum by a time discontinuity can be generalized to an arbitrary initial state $|\phi_1\rangle \equiv |n, n'\rangle_1$, not necessarily perfectly symmetric ($n \neq n'$). The probability of observing a given final state $|\phi_2\rangle \equiv |m, m'\rangle_2$ is determined by

$$p(m, m') =_2 \langle m, m'|\phi_1\rangle\langle\phi_1|m, m'\rangle_2. \tag{8.96}$$

In order to remain as close as possible to our previous discussion of the classical model for time reflection, we will concentrate on initial completely asymmetric states, where $n \neq 0$ photons pre-exist in the mode $\vec{k}$ and no photons at all propagate in the opposite direction, which means that $n' = 0$. This corresponds to an initial photon beam propagating along $\vec{k}$ for $t < 0$. The initial photon state is $|\phi_1\rangle = |n, 0\rangle_1$.

Using equations (8.84, 8.92), we obtain

$$|n, 0\rangle_1 = \frac{1}{\sqrt{n!}}[a_1^+(\vec{k})]^n|0, 0\rangle_1$$

$$= \frac{1}{\sqrt{n!}}\sqrt{1 - (B/A)^2} \sum_{m=0}^{\infty} \left(\frac{B}{A}\right)^m [a_1^+(\vec{k})]^n|m, m\rangle_2. \tag{8.97}$$

We can now use equation (8.77) and, noting that the operators $a_2^+(\vec{k})$ and $a_2(-\vec{k})$ commute, we obtain, after a binomial expansion

$$[a_1^+(\vec{k})]^n = A^n \sum_{r=0}^{n} \frac{n!}{(n-r)!r!}(-1)^{n-r} \left(\frac{B}{A}\right)^r [a_2^+(\vec{k})]^{n-r}[a_2(-\vec{k})]^r. \tag{8.98}$$

Using this in equation (8.97), we arrive at

$$|n, 0\rangle_1 = \sum_{r=0}^{n} \sum_{s=0}^{\infty} b_{sr}(n)|n + s, s\rangle_2 \qquad (8.99)$$

where we have used the new index $s = m - r$, and the coefficients

$$b_{sr}(n) = \frac{1}{\sqrt{n!}} A^n \sqrt{1 - (B/A)^2} \frac{n!}{(n - r)!r!} (-1)^{n-r} \left(\frac{B}{A}\right)^{s+2r}. \qquad (8.100)$$

From this we can calculate the probability for observing at a time $t > 0$ a given photon state $|n + s, s\rangle_2$:

$$p(n, s) =_2 \langle n + s, s|n, 0\rangle_{11}\langle n, 0|n + s, s\rangle_2 = \left|\sum_{r=0}^{n} b_{sr}(n)\right|^2. \qquad (8.101)$$

This result shows that, after the occurrence of a sudden time change at $t = 0$ of the refractive index of an infinite dielectric medium, there is a probability $p(n, 0) \neq 1$ of observing a state $|n, 0\rangle_2$ with the same number of photons propagating with the same wavevector (but with a shifted frequency).

But there is also a finite probability $p(n, s) \neq 0$ of observing a number $s > 0$ of photons propagating in the opposite direction $-\vec{k}$. This gives us the quantum explanation for the effect of time reflection already described with the classical theory of chapter 6.

The above discussion can easily be generalized to include the case of coherent states. It is well known that these particular quantum states have the advantage of tending to a classical field in the limit of large occupation numbers.

By definition, a coherent state for a given field mode $\vec{k}$ is

$$|\alpha\rangle = e^{-|\alpha|^2/2} \sum_{n=0}^{\infty} \frac{\alpha^n}{\sqrt{n!}} |n_k\rangle. \qquad (8.102)$$

This means that we can, in our symmetric representation, define an initial coherent state as

$$|\alpha, 0\rangle_1 = e^{-|\alpha|^2/2} \sum_{n=0}^{\infty} \frac{\alpha^n}{\sqrt{n!}} |n, 0\rangle_1. \qquad (8.103)$$

Using equation (8.99) we can write this state vector in terms of the final states

$$|\alpha, 0\rangle_1 = e^{-|\alpha|^2/2} \sum_{n,m=0}^{\infty} \sum_{r=0}^{n} \frac{\alpha^n b_{sr}(n)}{\sqrt{n!}} |n + s, s\rangle_2. \qquad (8.104)$$

This expression shows that, in general, an initial coherent state will not lead to time-transmitted and time-reflected coherent states. This means that the time discontinuity does not preserve the classical-like properties of the initial photon state. However, it is also obvious that, for very large values of the initial occupation number $n \equiv n_k$ the transmitted state will be very similar to the initial one and close to a similar coherent state, propagating in a similar direction but with a shifted frequency. This problems would deserve a detailed numerical study [74].

Furthermore, the non-conservation of the classical-like properties of the initial states has to be related to the creation of squeezed states (states with no classical counterpart), which is also associated with the time discontinuity. This problem is not directly related to photon acceleration and will not be discussed here.

8.2.4   Conservation relations

We conclude the quantum theory of time refraction by briefly discussing the energy and momentum conservation relations. Let us first consider the total energy operator

$$W = \int w_k \frac{d\vec{k}}{(2\pi)^3} = \int \hbar\omega_k \left[a^+(\vec{k})a(\vec{k}) + \frac{1}{2} \right] \frac{d\vec{k}}{(2\pi)^3}. \tag{8.105}$$

For an initial state $|n, 0\rangle_1$, the expectation value of this operator is

$$\begin{aligned}
\langle W \rangle_1 &= {}_1\langle n, 0|W|n, 0\rangle_1 \\
&= \hbar\omega_1 \left(\langle n_k|a^+(\vec{k})a(\vec{k})|n_k\rangle + \langle 0|a^+(-\vec{k})a(-\vec{k})|0\rangle + 1 \right) \\
&= \hbar\omega_1(n_k + 1)
\end{aligned} \tag{8.106}$$

with $\omega_1 = |k|c/n_1$.

We have seen that the time discontinuity generates final states $|n + s, s\rangle_2$, with a finite probability $p(n, s)$. The expectation value for the energy operator for $t > 0$ will then be

$$\begin{aligned}
\langle W \rangle_2 &= {}_2\langle n + s, s|W|n + s, s\rangle_2 \\
&= \hbar\omega_2 \left(\langle n + s|a^+(\vec{k})a(\vec{k})|n + s\rangle + \langle s|a^+(-\vec{k})a(-\vec{k})|s\rangle + 1 \right) \\
&= \hbar\omega_2(n_k + 2s + 1)
\end{aligned} \tag{8.107}$$

with $\omega_2 = |k|c/n_2$.

This means that the energy variation introduced by the time discontinuity of the medium on the electromagnetic spectrum is

$$\Delta W = \langle W \rangle_2 - \langle W \rangle_1 = \hbar\Delta\omega n_k + \hbar\omega_2 2s. \tag{8.108}$$

We see that the energy is not conserved for two distinct reasons. The first one was already observed in the classical description and corresponds to the frequency shift $\Delta\omega = \omega_2 - \omega_1$ of the n_k photons initially existing in the medium. The second one is specific to the quantum nature of light and is due to the generation at $t = 0$ of oppositely propagating s pairs of photons.

Let us consider next the total momentum operator

$$P = \int p_k \frac{d\vec{k}}{(2\pi)^3} = \int \hbar \vec{k} \left[a^+(\vec{k})a(\vec{k}) + \frac{1}{2} \right] \frac{d\vec{k}}{(2\pi)^3}. \tag{8.109}$$

Using the same kind of approach we can calculate the initial and the final expectation values of this operator. The result is

$$\langle P \rangle_1 = \hbar \vec{k} n_k, \quad \langle P \rangle_2 = \hbar \vec{k}(n + s - s). \tag{8.110}$$

This means that the total momentum is conserved, as expected: $\Delta P = \langle P \rangle_2 - \langle P \rangle_1 = 0$.

8.3 Quantum theory of diffraction

Let us consider a sharp boundary between two stationary dielectric media with no dispersion. For simplicity, we assume that the media, with dielectric constants equal to ϵ_1 and ϵ_2, have a boundary at $x = 0$, and that the propagation is along the x-axis.

If photons with frequency ω and initial wavevector k_i are propagating in medium 1 and interact with this boundary, we can write for the associated electric field operator, valid in the semi-infinite region $x < 0$, and for a given polarization ($\lambda = 1$, or 2)

$$\vec{E}(x, t) = \vec{E}_i(x, t) + \vec{E}_r(x, t) \tag{8.111}$$

where the incident and the reflected field operators are

$$\vec{E}_i(x, t) = i \sqrt{\frac{\hbar \omega}{2\epsilon_1}} \left[a_1(k_i, t) e^{ik_i x} - a_1^+(k_i, t) e^{-ik_i x} \right] \vec{e}(k_i) \tag{8.112}$$

$$\vec{E}_r(x, t) = i \sqrt{\frac{\hbar \omega}{2\epsilon_1}} \left[a_1(k_r, t) e^{ik_r x} - a_1^+(k_r, t) e^{-ik_r x} \right] \vec{e}(k_r) \tag{8.113}$$

with $k_i = \omega n_1/c$ and $k_r = k_i$.

In the second medium ($x > 0$), we can define the electric field operator associated with the transmitted wave as

$$\vec{E}_t(x, t) = i \sqrt{\frac{\hbar \omega}{2\epsilon_2}} \left[a_2(k_t, t) e^{ik_t x} - a_2^+(k_t, t) e^{-ik_t x} \right] \vec{e}(k_t) \tag{8.114}$$

with $k_t = \omega n_2/c = k_i(n_2/n_1)$.

The corresponding magnetic field operators are determined by similar expressions:

$$\vec{B}_i(x, t) = \frac{k_i}{\omega} E_i(x, t)[\vec{e}_x \times \vec{e}(k_i)] \tag{8.115}$$

$$\vec{B}_r(x, t) = \frac{k_r}{\omega} E_r(x, t)[\vec{e}_x \times \vec{e}(k_r)] \tag{8.116}$$

and

$$\vec{B}_t(x, t) = \frac{k_t}{\omega} E_t(x, t)[\vec{e}_x \times \vec{e}(k_t)]. \tag{8.117}$$

The quantization of the electromagnetic field is based on the assumption that the field operators satisfy Maxwell's equations. This means that the boundary conditions for these operators have to be formally identical to those for the classical fields. We know, from classical theory, that the components of the electric and magnetic fields tangent to the boundary between the two media are continuous.

We can then write, for the operators

$$(\vec{E}_i + \vec{E}_r) \times \vec{e}_x = \vec{E}_t \times \vec{e}_x, \quad (\vec{B}_i + \vec{B}_r) \times \vec{e}_x = \vec{B}_t \times \vec{e}_x. \tag{8.118}$$

Using equation (8.117) and noting that the field is polarized in the perpendicular direction $\vec{e}(k_i) = \vec{e}(k_r) = \vec{e}(k_t)$, we obtain

$$E_i(0, t) + E_r(0, t) = E_t(0, t), \quad B_i(0, t) + B_r(0, t) = B_t(0, t). \tag{8.119}$$

At this point we notice that the time-dependent destruction and creation operators in equations (8.113, 8.114) are of the form

$$a(k, t) = a(k) e^{-i\omega t}, \quad a^+(k, t) = a^+(k) e^{i\omega t}. \tag{8.120}$$

Equating separately the terms with the same time dependence, we obtain

$$a_1(k_i) + a_1(-k_i) = \alpha^2 a_2(k_t)$$
$$a_1(k_i) - a_1(-k_i) = a_2(k_t) \tag{8.121}$$

with

$$\alpha = \sqrt{\frac{n_1}{n_2}} = \left(\frac{\epsilon_1}{\epsilon_2}\right)^{1/4}. \tag{8.122}$$

From this we obtain

$$a_2(k_t) = \frac{2}{1+\alpha^2} a_1(k_i), \quad a_1(-k_i) = -\frac{1-\alpha^2}{1+\alpha^2} a_1(k_i). \tag{8.123}$$

These expressions can be seen as the Fresnel formulae relating the incidence, the transmission and the reflection destruction operators. We can easily realize that the same expressions are valid for the creation operators. We can then derive from (8.123) a more familiar version of the Fresnel formulae, by multiplying them by $\exp(-i\omega t)$, and by adding their Hermitian conjugates. We therefore get

$$E_i(0, t) + E_r(0, t) = E_t(0, t), \quad E_i(0, t) - E_r(0, t) = \frac{1}{\alpha^2} E_t(0, t). \quad (8.124)$$

This is nothing but a different version of equation (8.119). From this, we get

$$R \equiv \frac{E_i(0, t)}{E_r(0, t)} = -\frac{1 - \alpha^2}{1 + \alpha^2}, \quad T \equiv \frac{E_t(0, t)}{E_i(0, t)} = \frac{2\alpha^2}{1 + \alpha^2}. \quad (8.125)$$

These operator relations are equivalent to equation (8.123), and they are formally identical to the Fresnel formulae for the classical fields. The same result could be obtained by using a different method [94].

For comparison with the time reflection case, it is interesting to consider the energy and momentum conservation relations. Let us start with the number operator associated with the incident photon states: $N_i \equiv a^+(k_i)a(k_i)$. By using equations (8.123) and noting that $a_2(k_t)$ and $a_1(-k_i)$ commute, we obtain

$$N_i = [a_1^+(-k_i) + a_2(k_t)][a_1(-k_i) + a_2(k_t)]$$
$$= a_1^+(-k_i)a_1(-k_i) + a_2^+(k_t)a_2(k_t) = N_r + N_t. \quad (8.126)$$

We see from here that, for a generic quantum state of the radiation field, the expectation value of the incident number operator N_i is equal to the sum of the expectation values of the reflected and transmitted photons: $n_i = n_r + n_t$. In addition, because the photons maintain their frequency upon reflection and refraction, the total energy is also conserved:

$$\Delta W = (W_r + W_t) - W_i = \hbar\omega(n_r + n_t) - \hbar\omega n_i = 0. \quad (8.127)$$

We see that photon creation from a vacuum around a space boundary cannot exist, in contrast with the case of a time boundary. On the other hand, the total momentum is not conserved, in agreement with the results of the classical theory:

$$\Delta P = (P_r + P_t) - P_i = \hbar(k_r n_r + k_t n_t) - \hbar k_i n_i$$
$$= \hbar(k_i + k_t)n_t = \hbar k_i \left(1 + \frac{1}{\alpha^2}\right) n_t. \quad (8.128)$$

Let us conclude with a comment on the more general case of a moving space boundary. It is clear from the above results that we can follow the approach used in the classical theory: first, we make a Lorentz transformation to the reference frame moving with the dielectric boundary. Second, we repeat the above quantization procedure in the moving frame and, finally, we make an inverse Lorentz transformation back to the rest frame. The result cannot be other than a relation between the incidence, reflection and transmission field operators identical to that obtained for the classical fields.

This leads us to the conclusion that the quantum theory of photon refraction at a moving boundary will not predict the creation of photons from a vacuum, because such an effect is already absent in the moving frame (or for a boundary at rest). This is in contrast with the quantum theory of section 8.2 which shows that such an effect is present for a time discontinuity, or equivalently, for a boundary moving with an infinite velocity.

Chapter 9

New developments

The theory of photon acceleration, as presented in its various versions in the present book, has recently been extended to new and exciting areas of physics. Two examples are given here.

One example is related to the new field of neutrino interactions with dense plasmas [13]. Here, the striking similarities between the photon and the neutrino dispersion relations in a plasma can be explored [68]. A considerable amount of theoretical work has already been performed by several authors, leading to an already quite coherent view of the collective neutrino–plasma interactions [106].

The other example concerns the coupling between photons and the gravitational field. Of particular interest is the attempt to describe photon acceleration processes associated with gravitational waves [72]. Given the extreme difficulty of this subject, these results have to be received with some caution, and can be considered as merely tentative. But, we firmly believe that these attempts can be very positive for the progress of knowledge in this yet quite unexplored area.

9.1 Neutrino–plasma physics

The neutrino interaction with very dense plasmas is of considerable importance in the early universe and during supernova explosions. The mechanisms for photon acceleration in non-stationary media can easily be transposed to neutrino–plasma physics.

It is well known that, in the presence of matter, the neutrino effective mass is changed due to the weak-current interaction. In particular, the charged current couples the electron neutrinos with the electrons existing in a dense plasma.

In order to describe this weak coupling we can use the dispersion relation of a neutrino in a plasma, relating its momentum $\vec{p}$ and its energy E [10, 119]:

$$(E - V)^2 - p^2 c^2 - m_\nu^2 c^4 = 0 \tag{9.1}$$

where m_ν is the (eventually existing) neutrino rest mass and V is an equivalent potential energy such that

$$V = \sqrt{2} G_F \frac{\epsilon_0 m}{e^2} \omega_p^2 \equiv g \omega_p^2. \tag{9.2}$$

Here G_F is the Fermi constant for weak interactions, e and m are the electron charge and mass and ϵ_0 is the vacuum permittivity. Because the coupling constant g is very small, we can assume, even for extremely dense plasmas, that the equivalent potential energy V is always much smaller than the total neutrino energy E.

We also know that the standard theory for electro-weak interactions assumes a zero value for the neutrino rest mass m_ν [88], but some recent observations of solar neutrinos point to the existence of a small rest mass [32]. For that reason, we opt here for retaining the rest mass in the neutrino dispersion relation. This relation can be rewritten in terms of the neutrino frequency ω and of the neutrino wavevector $\vec{k}$ (using $\hbar = 1$), for a space- and time-varying plasma, as

$$\omega(\vec{r}, \vec{k}, t) = \sqrt{k^2 c^2 + m_\nu^2 c^4 + g \omega_p^2 (\vec{r}, t)}. \tag{9.3}$$

This is indeed very similar to the photon dispersion relation. Here we can also use the ray equations for the neutrino field, which are formally identical to the equations for single photon trajectories. In the presence of electron plasma perturbations moving with a velocity $\vec{u}$, the electron plasma frequency is such that $\omega_p^2(\vec{r}, t) \equiv \omega_p^2(\vec{\eta})$, with $\vec{\eta} = \vec{r} - \vec{u}t$.

Following the procedure described in chapter 3, we can establish an invariant for the neutrino trajectories in the form $I = \omega - \vec{k} \cdot \vec{u}$. From this we conclude that the total energy exchange between the neutrino and the background plasma medium, when it crosses the moving plasma boundary (for instance, during the collapse of an exploding neutron star) is

$$\Delta E = g \omega_{p0}^2 \frac{\beta}{1 - \beta} \tag{9.4}$$

where $\beta = |u|/c$ and where we have supposed counter-propagation between the neutrino and the plasma front.

It should be noticed that, as before for the photon case, this process of energy exchange between the particle and its background plasma medium is linear and not resonant, which means that all the particles with the same initial energy will exchange the same amount of energy with the medium.

This simple approach is based on the classical description of the neutrino, equivalent to the geometric optics approximation for the photon field. This is clearly valid as long as the plasma space and timescales are much larger than those characterizing the neutrino state.

But we can do better and use a quantum mechanical description of this non-resonant interaction between the neutrinos and the plasma medium. For that purpose, we derive from the above dispersion relation (9.1) a Klein–Gordon wave equation by making the usual replacement:

$$\vec{p} \rightarrow -i\hbar \frac{\partial}{\partial \vec{r}}, \quad E \rightarrow i\hbar \frac{\partial}{\partial t}. \tag{9.5}$$

From this we get

$$\left(\hbar^2 c^2 \nabla^2 - m_\nu^2 c^4 - \hbar^2 \frac{\partial^2}{\partial t^2} \right) \psi = 2i\hbar V \frac{\partial \psi}{\partial t} \tag{9.6}$$

where ψ is the normalized wavefunction associated with the neutrino field.

We could improve our quantum description of the neutrino field by replacing this Klein–Gordon equation by a more adequate Dirac equation. But, the present formulation stays valid as long as we neglect the coupling between different helicity states.

We are interested in the case where the effective potential $V(\vec{r}, t)$ is associated with a plasma perturbation moving with a velocity $\vec{u}$. If the perturbation is nearly uniform in the plane perpendicular to its velocity, we can choose the x-axis as parallel to $\vec{u}$ and reduce the wave equation with its one-dimensional version:

$$\left(\frac{\partial^2}{\partial x^2} - \frac{1}{c^2} \frac{\partial^2}{\partial t^2} - \frac{m_\nu^2 c^2}{\hbar^2} \right) \psi(x, t) = \frac{2i}{\hbar c^2} V(x - ut) \frac{\partial}{\partial t} \psi(x, t). \tag{9.7}$$

Let us now use the D'Alembert transformation of variables:

$$\xi = x - ut, \quad \eta = x + ut. \tag{9.8}$$

We can easily see that

$$\left(\frac{\partial^2}{\partial x^2} - \frac{1}{c^2} \frac{\partial^2}{\partial t^2} \right) = (1 - \beta^2) \left(\frac{\partial^2}{\partial \xi^2} + \frac{\partial^2}{\partial \eta^2} \right) + 2(1 + \beta)^2 \frac{\partial^2}{\partial \eta \partial \xi}. \tag{9.9}$$

When $u = c$, or $\beta = 1$, this reduces to the expression

$$\left(\frac{\partial^2}{\partial x^2} - \frac{1}{c^2} \frac{\partial^2}{\partial t^2} \right) = 4 \frac{\partial^2}{\partial \eta \partial \xi}. \tag{9.10}$$

When written in terms of the new variables, the wave equation (9.7) becomes

$$\left[(1 - \beta^2) \left(\frac{\partial^2}{\partial \xi^2} + \frac{\partial^2}{\partial \eta^2} \right) + 2(1 + \beta)^2 \frac{\partial^2}{\partial \eta \partial \xi} - \frac{m_\nu^2 c^2}{\hbar^2} \right] \psi(\xi, \eta)$$

$$= \frac{2i\beta}{\hbar c} V(\xi) \left(\frac{\partial}{\partial \eta} - \frac{\partial}{\partial \xi} \right) \psi(\xi, \eta). \tag{9.11}$$

Because the potential V is independent of the variable η, we can introduce a Fourier transformation in that variable, such that

$$\psi(\xi, \eta) = \int \psi_q(\xi) \, e^{iq\eta} \frac{dq}{2\pi}. \tag{9.12}$$

This means that we can reduce the wave equation to

$$\left[(1 - \beta^2)\left(\frac{\partial^2}{\partial\xi^2} - q^2 \right) + 2iq(1 + \beta)^2 \frac{\partial}{\partial\xi} - \frac{m_\nu^2 c^2}{\hbar^2} \right] \psi_q(\xi)$$

$$= \frac{2i\beta}{\hbar c} V(\xi) \left(iq - \frac{\partial}{\partial\xi} \right) \psi_q(\xi). \tag{9.13}$$

In order to understand the meaning and to test the validity of this form of wave equation, let us first briefly consider the particular case of neutrino motion in a vacuum: $V(\xi) = 0$. In this trivial case, we can perform a Fourier transformation in the remaining variable, ξ, of the form

$$\psi_q(\xi) = \int \psi_{qp} \, e^{ip\xi} \frac{dp}{2\pi}. \tag{9.14}$$

The wave equation then leads to a neutrino dispersion relation of the form

$$(1 - \beta^2)(p^2 + q^2) + 2(1 + \beta^2)pq + \frac{m_\nu^2 c^2}{\hbar^2} = 0. \tag{9.15}$$

This strange expression has to be equivalent to the dispersion relation (9.3), with $\omega_p = 0$, which means it is equivalent to

$$\omega^2 - k^2 c^2 - \frac{m_\nu^2 c^4}{\hbar} = 0. \tag{9.16}$$

In order to be convinced of such an equivalence, we notice that

$$\frac{\partial}{\partial x} = \frac{\partial}{\partial\xi} + \frac{\partial}{\partial\eta}, \quad \frac{\partial}{\partial t} = u\frac{\partial}{\partial\eta} - \frac{\partial}{\partial\xi}. \tag{9.17}$$

This leads to the two relations

$$k = p + q, \quad \omega = u(p - q). \tag{9.18}$$

Using this in equation (9.15), we obtain equation (9.16). It means that this is indeed the neutrino dispersion relation in a vacuum, written in terms of the more complicated quantities p and q.

The physical meaning of these new variables is given by (9.18): their sum is the neutrino momentum and their difference is its energy divided by its velocity.

If the neutrino rest mass was made equal to zero, these two quantities, k and ω/u, would be identical.

We can now return to the non-trivial case of a slowly varying potential, $V(\xi) \neq 0$. Here, we can try a WKB solution for the wave equation (9.13) of the form

$$\psi_q(\xi) = \psi_{q0} \exp \left(i \int^\xi p(\xi') \, d\xi' \right). \tag{9.19}$$

With this, we can obtain a dispersion relation, which is locally valid in ξ:

$$(1-\beta^2) \left[p(\xi)^2 + q^2 \right] + 2(1+\beta^2) p(\xi) q + \frac{m_\nu^2 c^2}{\hbar^2} = \frac{2\beta}{\hbar c} V(\xi) [q - p(\xi)]. \tag{9.20}$$

For this type of solution to be valid, equations (9.18) have to be replaced by similar ones, valid only locally, and which can be written as

$$2q = k(\xi) - \frac{\omega(\xi)}{u}, \quad 2p(\xi) = k(\xi) + \frac{\omega(\xi)}{u}. \tag{9.21}$$

Here, the quantities $k(\xi)$ and $\omega(\xi)$ represent the local values of the wavenumber and frequency for a given neutrino state characterized by the quantity q. This quantity remains constant over the entire trajectory. Due to the existence of such an invariant (which also appeared in the classical theory), we can easily calculate the change of the neutrino energy when it interacts with a non-stationary plasma background, in the same way as we have established the photon frequency shift in a non-stationary optical medium.

In particular, if we assume that the plasma potential tends to zero at infinity, $V(\xi \to \infty) = 0$, and if in the other side of the discontinuity at $\xi = 0$ its value is V_0, then we can derive from the invariance of q an energy shift equal to the one given by equation (9.4). But, apart from confirming the classical results for the energy variation, the quantum description can also reveal a qualitative new effect, namely the possibility of quantum reflection at the moving plasma boundary.

Using equations (9.21), we can easily obtain a relation between the energy of the incident and the reflected energy states, $E_i = \hbar\omega_i$ and $E_r = \hbar\omega_r$:

$$-2qu = k_i u + \omega_i = -k_r u + \omega_r. \tag{9.22}$$

Assuming that we nearly have $kc = \omega$, we get from this the relativistic mirror relation $\omega_r = \omega_i(1 + \beta)/(1 - \beta)$. According to the same equations (9.21), the values of the quantum number p associated with the incident and the reflected wave solution, for the same value of q, are different:

$$2p_i u = k_i u - \omega_i \simeq -\omega_i(1 - \beta), \quad 2p_r u = -k_r u - \omega_r \simeq -\omega_r(1 + \beta). \tag{9.23}$$

Using the above relativistic mirror relation, we obtain

$$p_r \simeq p_i \frac{(1 + \beta)^2}{(1 - \beta)^2}. \tag{9.24}$$

The probability amplitudes for the reflected neutrino state can also be derived from the wave equation, by using a mode coupling approach. If we assume wave function solutions of the form

$$\psi_q(\xi) = \sum_{j=1,2} \psi_{qj}(\xi)\, e^{i \int^\xi p_j(\xi')\, d\xi'} \tag{9.25}$$

where $\psi_{qj}(\xi)$ are slowly varying amplitudes such that

$$\frac{\partial \psi_{qj}}{\partial \xi} \ll p_j \psi_{qj}, \tag{9.26}$$

this means that we can write

$$\frac{\partial^2}{\partial \xi^2} \psi_{qj} \simeq -p_j^2 \psi_{qj} + i p_j \frac{\partial \psi_{qj}}{\partial \xi}. \tag{9.27}$$

Using the solution (9.25) in the wave equation (9.13), and assuming that the local dispersion relation (9.20) stays valid for both the incident and the reflected particle state, we obtain

$$\sum_j \left[(1 - \beta^2) p_j(\xi) \right] \frac{\partial \psi_{qj}}{\partial \xi}\, e^{i \int^\xi p_j(\xi')\, d\xi'}$$

$$\simeq \frac{2\beta}{\hbar c} V(\xi) \sum_{l \neq j} [q - p_l(\xi)]\, \psi_{ql}\, e^{i \int^\xi p_l(\xi')\, d\xi'}. \tag{9.28}$$

We know that the neutrino–plasma interaction is very weak, even for very dense plasmas, due to the extremely small value of the Fermi constant G_F. This means that we can assume, in quite general conditions, that the incident mode ψ_{q1} is dominant, and that its amplitude is only slightly perturbed by the interaction with the moving plasma discontinuity.

We can then write the evolution equation for the reflected mode in the parametric form

$$\frac{\partial \psi_{q2}}{\partial \xi} = w(\xi) \psi_{q1} \tag{9.29}$$

where the coupling coefficient is determined by

$$w(\xi) = \frac{2}{\hbar c} \frac{\beta}{(1 - \beta)^2} \frac{V(\xi)}{p_1(\xi)}\, e^{i \int^\xi (p_1 - p_2)\, d\xi'}. \tag{9.30}$$

This expression shows that the coupling between the initial and the reflected state is proportional to the velocity of the moving plasma perturbation, or equivalently to the value of β. It is also proportional to the Fourier component of the potential $V(\xi)$ for $p = p_1 - p_2$.

This can be seen by integrating equation (9.29) and assuming that the initial value of the reflected state amplitude is zero (or $\psi_{q2}(\xi_0) = 0$):

$$\psi_{q2}(\xi) = \psi_{q1} \int_{\xi_0}^{\xi} w(\xi')\, d\xi'. \tag{9.31}$$

We can further explore the analogy between the neutrino and the photon coupling with a non-stationary background plasma by developing a kinetic theory for the neutrino gas present in the medium. The result is a neutrino kinetic equation, or neutrino fluid equations, similar to those derived in chapter 4 for the photon gas [112].

We should also keep in mind that the neutrino gas will react back on the plasma electrons. This can be described by a neutrino pressure term, similar to the radiation pressure considered before, which has to be added to the electron equations of motion. Associated with this pressure term we can likewise define an equivalent electric charge for the neutrino gas [69, 83, 85, 113].

The result of this mutual coupling between the neutrino and the electron gas in a plasma leads to the possibility of neutrino Landau damping of relativistic electron plasma oscillations [73], and to the possibility of transferring a considerable fraction of the energy of a neutrino beam into the plasma, by exciting electron plasma waves in the medium [102]. Other surprising results arising from the study of this collective neutrino plasma are the generation of magnetic fields and inhomogeneities in the early stages of the universe [105].

9.2 Photons in a gravitational field

The large variety of effects leading to a frequency shift (or to a spectral change) of the electromagnetic radiation, which we have classified under the name of photon acceleration, present some analogies with the well-known gravitational frequency shift occurring when photons escape from regions of strong gravitational fields. Even if the gravitational effects are described by a completely different theoretical approach, one can question the possible connections between these gravitational frequency shifts and our concept of photon acceleration.

Such a question is relevant, not only because it can lead to a deeper and more global view of photon dynamics, but also because it is known that, at least in some limit, a gravitational field can be adequately described as a dielectric medium in a flat space [55, 126]. We can then make the bridge between the photon propagation in a dielectric medium (in the absence of a gravitational field) as described in this book, and the photon propagation in the presence of a gravitational field (and in the absence of a medium).

9.2.1 Gravitational redshift

We show first that the Hamiltonian ray theory can be easily extended to include the influence of the gravitational fields. This can be done by choosing a convenient definition for the photon frequency, as explained below. This new theoretical approach will then provide an alternative derivation of the gravitational redshift. Once this question is clarified, we can eventually use a kinetic photon theory in a gravitational field, in the same way as we did for plasmas and for other dielectric media.

Let us consider an electromagnetic wavepacket characterized by the four-vector $k^i \equiv (\omega/c, \vec{k})$, where ω is the mean frequency and $\vec{k}$ the mean wavevector. We know that, in a vacuum and in the absence of a gravitational field, the absolute value of this four-vector vanishes:

$$k^i k_i = 0. \tag{9.32}$$

If a wave is propagating in a slowly varying gravitational field we can introduce an eikonal ψ, such that:

$$k_i = \partial \psi / \partial x^i. \tag{9.33}$$

If we replace this in the above condition for the absolute value of k^i we obtain the eikonal equation [55]

$$g^{ik} \frac{\partial \psi}{\partial x^i} \frac{\partial \psi}{\partial x^k} = 0 \tag{9.34}$$

where g^{ik} are the components of the metric tensor.

We will illustrate our ideas by using a very simple metric tensor such that, for an arbitrary but constant gravitational field, the spatial coordinate system is Cartesian at a given point and at a given time: $g^{ik} = -\delta_{ik}$, where the indices i and k are supposed to take the values 1, 2 and 3, corresponding to the space coordinates. As concerns the time coordinates, we use $g^{00} \neq 1$. More realistic (but also more complicated) metric tensors will be discussed at the end of this section.

If we define the frequency of the electromagnetric wave ω as the derivative of the eikonal function ψ with respect to the time variable

$$\omega = -c \frac{\partial \psi}{\partial x^0}, \tag{9.35}$$

we obtain from equation (9.34) the following dispersion relation:

$$\omega^2 = k^2 c^2 g_{00} \tag{9.36}$$

where we noticed that $g^{00} = 1/g_{00}$.

The same result could be obtained by noting [55, 126] that the influence of a (quasi-static) gravitational field on the electromagnetic wave propagation is equivalent to the change of the dielectric constant of a vacuum from ϵ_0 to $\epsilon_0/\sqrt{g_{00}}$, and a similar change of the magnetic permeability of a vacuum from μ_0 to $\mu_0/\sqrt{g_{00}}$.

Introducing a scalar potential V, such that $g_{00} = 1 + 2V/c^2$, we can rewrite this photon dispersion relation as

$$\omega = kc\sqrt{1 + 2\frac{V}{c^2}} \simeq kc + \frac{k}{c}V. \qquad (9.37)$$

We know that the frequency $\omega = \omega(\vec{r}, \vec{k}, t)$ can be used as the photon Hamiltonian for the ray equations written in the canonical form

$$\frac{d\vec{r}}{dt} = \frac{\partial \omega}{\partial \vec{k}} \simeq \left(c + \frac{V}{c}\right)\frac{\vec{k}}{k}$$

$$\frac{d\vec{k}}{dt} = -\frac{\partial \omega}{\partial \vec{r}} \simeq -\frac{k}{c}\frac{\partial V}{\partial \vec{r}}. \qquad (9.38)$$

For the total time derivative of the Hamiltonian function, we can also write

$$\frac{d\omega}{dt} = \frac{\partial \omega}{\partial t} \simeq \frac{k}{c}\frac{\partial V}{\partial t}. \qquad (9.39)$$

This equation shows that, for a static gravitational field, the frequency ω, as defined by equation (9.35), is a constant of motion. This will be discussed in more detail below.

If the potential V is due to a single massive object (for instance a star), of mass M and radius R, we can write $V(\vec{r}) = -GM/r$, for $r \geq R$. One photon emitted at the surface of that star will have the initial wavenumber k_1 such that $\omega = k_1 c - (k_1/c)GM/R$. This photon will be observed on earth with a wavenumber k_2, such that $(r \to \infty)$ $\omega = k_2 c$.

We obtain from these two expressions of ω the following change in the wavenumber:

$$\Delta k = k_2 - k_1 = -(k_1/c)GM/R < 0. \qquad (9.40)$$

This means that the observed wavelength observed on earth will be larger than the one emitted at the star surface ($\Delta\lambda > 0$). This is the well-known gravitational redshift.

In the present formulation, we could be led to the conclusion that the gravitational redshift appears only as a shift in wavelength, and not a shift in frequency, because the photon frequency ω, as defined by equation (9.35), is an invariant. However, this apparent contradiction with the conventional description of the gravitational redshift disappears if we notice that we are using a different definition for the photon frequency. It is well known that the conventional view is based on the photon 'proper frequency' ω_τ, which is defined as the derivative of the eikonal with respect to the proper time τ (notice that this is the observer proper time and not the photon proper time):

$$\omega_\tau = -\frac{\partial \psi}{\partial \tau} = -\frac{\partial \psi}{\partial x^0}\frac{\partial x^0}{\partial \tau} = \frac{\omega}{\sqrt{g_{00}}}. \qquad (9.41)$$

With this local definition of frequency we get $\omega_\tau = kc$ everywhere, meaning that $\Delta\omega_\tau = \Delta kc$. In this way, the gravitational redshift becomes a frequency shift as well.

We propose to use the universal time ω, instead of the proper time ω_τ, because it coincides with the Hamiltonian function appearing in the photon canonical equations (9.38). In addition, as discussed below, these equations can easily be extended to other situations and used to describe several other important effects, such as the photon bending in the vicinity of a star or the photon acceleration by a gravitational wavepacket.

9.2.2   Gravitational lens

Let us first discuss a gravitational lens, or photon bending near a massive star. The massive astrophysical objects that can produce important light bending are also very often surrounded by dense and warm plasmas. It is then useful to study the photon equations of motion in a plasma and in a gravitational field. In order to keep the discussion formally simple, we include the plasma dispersive effects but neglect the influence of static magnetic fields.

For $V = 0$, the dispersion relation for photons in a non-magnetized plasma is simply: $\omega_\tau^2 \equiv \omega^2 = k^2c^2 + \omega_p^2$, where ω_p is the electron plasma frequency. If a static gravitational field is also present, we have $V \neq 0$, and, using the above metric transformation, we obtain $\omega^2 = [k^2c^2 + \omega_p^2]g_{00}$, or more explicitly

$$\omega = \sqrt{(k^2c^2 + \omega_p^2)\left(1 + 2\frac{V}{c^2}\right)}. \tag{9.42}$$

Let us again assume that the static gravitational field is related to a star of mass M and radius R. We can then replace $V(r) = -GM/r$, for $r > R$ in this dispersion relation and use it as the appropriate Hamiltonian function for the photon canonical equations

$$\omega \equiv \omega(\vec{r}, \vec{k}) = \sqrt{(k^2c^2 + \omega_p^2)\left(1 - \frac{\alpha}{r}\right)} \tag{9.43}$$

with $\alpha = 2GM/c^2$.

In general, the plasma density will decay exponentially away from the surface of the star ($r > R$), according to a Boltzmann distribution. This means that we can rewrite it as

$$\omega_p^2 = \omega_{p0}^2 e^{-mV(r)/T} \simeq \omega_{p0}^2\left(1 - \frac{\beta}{r}\right) \tag{9.44}$$

where T is the plasma temperature, $\beta = mGM/T$ and the approximate expression is valid for a warm plasma such that $T \gg mV(r)$.

The photon equations of motions (9.38) can now be written as

$$\frac{d\vec{r}}{dt} = \frac{c^2}{\omega}\left(1 - \frac{\alpha}{r}\right)\vec{k}$$

$$\frac{d\vec{k}}{dt} = -\frac{\partial\omega}{\partial r}\vec{e}_r. \tag{9.45}$$

In the second of these equations we have used spherical coordinates $\vec{r} = (r, \theta, \phi)$. For motion in the plane $\phi = $ const, we can write

$$\vec{k} = k_r\vec{e}_r + (L/r)\vec{e}_\theta \tag{9.46}$$

where $L = $ const is the angular momentum of the photon trajectory. Assuming that along the photon trajectory we always have $(\alpha/r) \ll 1$ and neglecting the term in $1/r^2$, we obtain an approximate expression for the Hamiltonian:

$$\omega \simeq c\sqrt{k_r^2 + \frac{L^2}{r^2} + m_{\text{eff}}^2 c^2} - \frac{a}{r}. \tag{9.47}$$

Here we have used the photon equivalent mass in a plasma: $m_{\text{eff}} = \omega_{\text{p0}}/c^2$. We have also introduced the new parameter

$$a = \alpha\omega\left[1 - \frac{\omega_{\text{p0}}^2}{\omega^2}\left(1 - \frac{\beta}{\alpha}\right)\right]. \tag{9.48}$$

This approximate photon Hamiltonian is formally identical to that of a relativistic particle with mass m_{eff}, moving in a Coulomb field. The corresponding trajectories are well known from the textbooks [55] and it is not necessary to state them here.

What is important to notice is that this description of photon motion around a massive object generalizes the usual description of gravitational bending for the case of a plasma in a gravitational field. Notice also that diffraction due to plasma inhomogeneities around a star can dominate over purely gravitational bending if $\beta \ll \alpha$. In this case we have $a \simeq \beta\omega_{\text{p0}}^2/\omega$. In the absence of a plasma, we recover the well-known results concerning light bending by a star.

9.2.3 Interaction of photons with gravitational waves

It is already quite surprising and rewarding that we could rederive, from our simple photon equations of motion, the gravitational redshift and the gravitational lens effects, and moreover, that we could easily include the plasma refraction effects. The inclusion of plasma effects is quite interesting because stars or other massive objects are usually surrounded by warm and dense plasmas which can eventually influence the total lensing effect.

Let us now consider a much less obvious case, where the static gravitational field of the previous two cases is replaced by a time-dependent gravitational field. This is, in general, a difficult problem because in order to solve it, we have to make explicit use of the Einstein equations of general relativity.

However, if this time-varying field is due to an infinitesimal gravitational wavepacket, propagating in an otherwise flat space–time, we can still define a universal frequency and use the dielectric description of gravitation as an acceptably good approximation [126]. This means that we can use, as a first-order approximation, a flat space where the spatial part of the metric tensor is still Euclidian and the temporal metric component g_{00} becomes a function of time.

Let us neglect plasma effects and consider the propagation of a photon in a vacuum. Equation (9.36) shows that the universal photon frequency ω is no longer a constant, and the same will happen to the proper photon frequency ω_r. We can then say that, for a time variation g_{00}, photon acceleration (or energization) by the gravitational field can eventually occur.

In order to derive explicit results, let us use the following scalar potential:

$$V(\vec{r}, t) = V_0 + \tilde{V}(\vec{r} - \vec{v}_f t). \tag{9.49}$$

Here $\vec{v}_f$ is the velocity of the infinitesimal gravitational field perturbation and $\tilde{V}$ its amplitude. In analogy with several other similar occasions, we can introduce a canonical transformation from $(\vec{r}, \vec{k})$ to a new pair of variables $(\vec{\eta}, \vec{p})$, such that $\vec{\eta} = \vec{r} - \vec{v}_f t$ and $\vec{p} = \vec{k}$.

The photon equations of motion become

$$\begin{aligned}
\frac{\mathrm{d}\vec{\eta}}{\mathrm{d}t} &= \frac{\partial \omega'}{\partial \vec{p}} \\
\frac{\mathrm{d}\vec{p}}{\mathrm{d}t} &= -\frac{\partial \omega'}{\partial \vec{\eta}}.
\end{aligned} \tag{9.50}$$

The new Hamiltonian associated with these canonical equations is

$$\begin{aligned}
\omega'(\vec{\eta}, \vec{p}) &= \omega(\vec{\eta}) - \vec{v}_f \cdot \vec{p} \\
&= \left\{ (p^2 c^2 + \omega_p^2) \left[1 + 2\frac{V_0}{c^2} + 2\frac{\tilde{V}(\vec{\eta})}{c^2} \right] \right\}^{1/2} - \vec{v}_f \cdot \vec{p}.
\end{aligned} \tag{9.51}$$

Several physical configurations can be studied with these Hamiltonian formulations and all lead, even for very small moving gravitational perturbations, to strong photon acceleration. This is due to the nearly resonant character of the photon–gravitational-wave interaction because of their close group and phase velocities. Moreover, such an interaction takes place over extremely large distances, which enhances the process of photon acceleration.

In order to illustrate this general feature, let us consider two different situations. In the first one we assume that, near some radiating source (for instance a collapsing star), a gravitational shock wave starts to form before being radiated away, and interacts with the surrounding photons. Strictly speaking, this first example is not quite compatible with our assumption of a small gravitational perturbation travelling in a flat space. However, we will keep it here as a first and illustrative attempt to formulate a new problem that will be interesting for a future and more accurate solution.

We will retain the flat space approximation and describe the shock front in the following simple form of a potential perturbation:

$$\tilde{V}(\vec{\eta}) = \frac{\tilde{V}}{2}[1 + \tanh(\vec{k}_f \cdot \vec{\eta})]. \tag{9.52}$$

Here $\vec{k}_f$ determines the scale of the moving shock front. The photon equations of motion (9.50) could be solved numerically. However, because we are concerned with order of magnitude estimates, based on a very rough description, it is more interesting to extract some results from the invariance of ω'. This will allow us to discuss the relevant properties of the photon motion.

If the photons are emitted at some point $\eta \ll 0$, their initial frequency ω_1 is such that

$$\omega' = \omega_1 - \vec{v}_f \cdot \vec{k}_1 \tag{9.53}$$

where we assume that the shock front velocity $\vec{v}_f$ and the initial photon wavevector $\vec{k}_1$ are nearly parallel.

After crossing the entire shock front, the photons will acquire a final frequency ω_2 and a final wavevector $\vec{k}_2$, such that

$$\omega' = \omega_2 - \vec{v}_f \cdot \vec{k}_2. \tag{9.54}$$

Equating these two different expressions for the invariant ω', we obtain

$$\omega_2 = \omega_1 \frac{1 - \beta(1 - \delta)}{1 - \beta(\cos\theta_2/\cos\theta_1)(1 - \delta + \tilde{V}/c^2)} \tag{9.55}$$

where θ_1 and θ_2 are the initial and final angles between the photon wavevector and the front velocity, and where we have introduced the following auxiliary parameters:

$$\beta = \frac{v_f}{c}\cos\theta_1, \quad \delta = \frac{V_0}{c^2} + \frac{\omega_p^2}{2k_1^2 c^2}. \tag{9.56}$$

In order to have an estimate of the total frequency up-shift, let us assume that $\theta_1 = 0$ and that $\beta \sim 1$. We obtain

$$\Delta\omega = \omega_2 - \omega_1 = \omega_1 \frac{\tilde{V}}{c^2}\frac{\beta}{1 - \beta}. \tag{9.57}$$

For a perturbation $\tilde{V}/c \simeq 10^{-1}$ and $\epsilon = 1 - \beta \simeq 10^{-8}$, this leads to $\omega_2 \simeq \omega_1 \times 10^7$, which means that the background infrared or visible photons existing near the exploding star can be accelerated up to the gamma ray frequency range. Notice that the time necessary for the photon to cross the front will be, in this case, equal to $\tau = (2/k_f)$ seconds, if k_f is given in cm^{-1}.

As a second example, let us consider photons interacting in a vacuum with an infinitesimal gravitational wavepacket. In contrast with the previous example this is now well inside the domain of validity of the dielectric model for the gravitational field. Because we are considering propagation in a vacuum, we use $\omega_p = 0$ and write

$$\tilde{V} = \bar{V} \cos(\vec{k}_f \cdot \vec{\eta}). \tag{9.58}$$

The photon frequency shift will occur when it travels from a region of gravitational wave maximum to a wave minimum. In this case, the same kind of analysis leads to a maximum frequency shift which is twice the value given by equation (9.57).

But now we have a much smaller field perturbation ($\bar{V}/c^2 \ll 1$) and, because the photons and the gravitational wave travel almost exactly at the same speed, we have to assume that the angles θ_1 and θ_2 are small but not exactly zero. Otherwise, the total frequency shift would be zero.

We can then write

$$\Delta\omega \simeq \omega_1 \frac{4}{\theta_2^2} \frac{\bar{V}}{c^2}. \tag{9.59}$$

For an extremely small perturbation $\bar{V}/c^2 \simeq 10^{-9}$ we still obtain photon accelerations up to the gamma ray energies $\omega_2 \simeq 4\omega_1 \times 10^7$ for a very small angle between the photon wavevectors and the gravitational wavevechtor: $\theta_2 \simeq 10^{-8}$ radians. The distance necessary for the photons to travel from a maximum to a minimum of the gravitational wavepacket will be $d \simeq k_f^{-1} \times 10^{-2}$ parsecs, which is quite small on a cosmological scale.

We should point out that in these estimates we have neglected the focusing effects associated with the difference between the initial and the final angles, θ_1 and θ_2. However these two angles can easily be related, due to the existence of a second invariant for the photon trajectories, as discussed in chapter 3. This invariant states the conservation of the transverse photon wavevector, and can be written as $k_1 \sin\theta_1 = k_2 \sin\theta_2$.

The result of this two-dimensional effect is to focus the photon trajectories in the direction of the gravitational wave propagation, leading to a decrease in the local angles θ and an increase in the effective β. This is due to the fact that the photon acceleration process increases the parallel photon wavevector, while the transverse one is kept constant. It then enhances the acceleration process (at the expense of a larger interaction distance) and broadens the region of photon phase space with a significant frequency up-shift.

Notice that the value of $\beta = 1$ is never exactly attained by these trajectories. This means that the exactly resonant condition between the photons and the gravitational waves, for which no acceleration exists, is an ensemble of zero measure in the photon phase space and, quite fortunately, is physically irrelevant to this problem.

It should be noted that, for arbitrary field amplitudes, the wave solutions of the Einstein equations generally imply that, apart from g_{00}, other time-dependent components of the metric tensor should also be included. This is equivalent to saying that, in general, a universal frequency cannot be defined. This will lead to the necessity of using a different dispersion relation.

9.2.4 Other metric solutions

Let us briefly show how the above description of photon motion in a gravitational field can be improved, by using more accurate metric tensors. First of all, we should notice that the general dispersion relation in a vacuum (9.34) can be explicitly written as

$$g^{00}\omega^2 - 2g^{0\alpha}\omega k_\alpha c + g^{\alpha\beta}k_\alpha k_\beta c^2 = 0 \tag{9.60}$$

where $\alpha = 1, 2, 3$, $\omega = -ck_0$, and we have used the symmetry $g^{0\alpha} = g^{\alpha 0}$.

If the vacuum was replaced by a plasma medium, we would have to replace the zero, in the right-hand side of this equation, by ω_p^2, due to the existence of a cut-off frequency. We can then use the explicit form of the Hamiltonian function $\omega = \omega(k_\alpha, x^\alpha, t)$, where $t = x^0/c$ in the photon canonical equations (2.97).

These equations (9.60, 2.97) are valid in quite general conditions. Let us give some examples of metric solutions, which can be physically more accurate than the simple example used above.

For instance, if we have a weak gravitational field, the interval is determined by

$$ds^2 = g_{ij}\, dx^i\, dx^j = \left(1 + \frac{V}{c^2}\right)c^2\, dt^2 - \left(1 - \frac{V}{c^2}\right)dl^2. \tag{9.61}$$

If the field is created by a star with mass M, we have $V = -2GM/r$, as stated above. The corresponding metric tensor will have components $g_{00} = (1 + V/c^2)$ and $g_{\alpha\alpha} = -(1 - V/c^2)$, for $\alpha = 1, 2, 3$. The dispersion relation in a plasma will be given by

$$\omega^2 = \left(\omega_p^2 - \frac{k^2 c^2}{g_{\alpha\alpha}}\right)g_{00}. \tag{9.62}$$

This is not very much different from equation (9.42). Let us now consider a Schwarzschild type of metric, which is valid for a non-rotating spherical distribution of mass. The interval can be written in spherical coordinates as

$$ds^2 = A(r)c^2\, dt^2 - B(r)\, dr^2 - r^2\, d\theta^2 - r^2\sin^2\theta\, d\xi^2 \tag{9.63}$$

where $A(r)$ and $B(r)$ are functions of the radial coordinate r.

The photon dispersion equation can now be written as

$$\omega^2 - \frac{\omega_p^2}{B(r)} = \frac{A(r)}{B(r)}k_r^2 + \frac{A(r)}{r^2}k_\perp^2 \qquad (9.64)$$

with

$$k_\perp^2 = k_\theta^2 + \frac{k_\xi^2}{\sin^2\theta}. \qquad (9.65)$$

This expression shows a clear asymmetry between the radial and the perpendicular propagation.

As a final example of a photon dispersion relation in a gravitational field, let us consider the metric solution of a gravitational wavepacket of the form

$$ds^2 = c^2 dt^2 - dl^2 + f(x,t)(c\,dt - dx)^2. \qquad (9.66)$$

We can use

$$f(x,t) = a\cos(k_0 x^0 + k_1 x^1) = a\cos(qx - \Omega t) \qquad (9.67)$$

where a and Ω are the gravitational wave amplitude and frequency.

It can easily be shown that the photon dispersion relation in a vacuum takes the form

$$(1-f)\left(\frac{\omega}{c}\right)^2 + 2f\left(\frac{\omega}{c}\right)k_\| - (1+f)k_\|^2 - k_\perp^2 = 0. \qquad (9.68)$$

Here, we have used the photon wavenumbers parallel and perpendicular to the direction of the gravitational wave propagation: $k_\| = k_1$ and $k_\perp^2 = k_2^2 + k_3^2$. Solving for ω, we obtain

$$\omega = -\frac{f}{1-f}k_\| c \pm c\sqrt{\frac{k_\|^2}{(1-f)^2} + \frac{k_\perp^2}{1-f}}. \qquad (9.69)$$

From here we can see that, for photon propagation perpendicular to the direction of the gravitational wave, the dispersion relation is simply given by $\omega = k_\perp c/(1-f)$, and for parallel photon propagation, we are reduced to the case of pure vacuum $\omega = k_\| c$. This last expression shows that photons with an exactly parallel motion cannot be accelerated by the gravitational wave, as already stated before.

These various examples of metric solutions, and of the corresponding photon dispersion relations, show that the exchange of energy between the electromagnetic (or photon) field and the gravitational field is possible in a variety of situations. For a stationary gravitational field, our approach leads to a new and simple derivation of the well-known gravitational redshift and gravitational lens effects.

For moving gravitational field perturbations, we have shown that the photons can resonantly interact (in a plasma or in a vacuum) with gravitational shocks and with gravitational wavepackets. According to our rough estimates, this interaction will eventually lead to very high frequency up-shifts, over distances well inside the cosmological constraints.

This could eventually provide a natural explanation for the recently observed gamma ray bursts [75]. However, our estimates need to be validated by more credible metric solutions and more accurate numerical calculations.

Nevertheless, it results from our equations that such an interaction is universal, in the sense that it affects every photon moving in a gravitational field. We also believe that, apart from its specific astrophysical implications, our simple approach points to a new and fundamental coupling mechanism between the electromagnetic and the gravitational waves which should be investigated in the future.

Actually, it has been known for quite some time that a gravitational wave can be Landau damped by a background photon gas [18], but the physical consequences of such damping, which is the statistical counterpart of the acceleration of photons by a gravitational wave, has not yet been explored.

9.3   Mean field acceleration processes

This section contains a few concluding remarks. We have started with very simple ideas concerning the processes of time refraction and time reflection, which correspond to a natural extension of the familiar concepts of refraction and reflection into the space–time domain.

The frequency shift resulting from these basic processes is what we call photon acceleration. We have shown that it can occur in a large variety of physical situations, in plasmas, in optical fibres or other optical media, and that it is more effective when the space–time disturbance of the medium travels with a velocity nearly equal to that of the photons.

Presently, a significant fraction of the physics community is still reluctant to talk about photon acceleration and prefers to use other terms such as frequency shift or phase modulation. This is not, in principle, a big problem, because what is important in physics is to have an accurate view of the physical processes, independent of the way they are known. However, the choice of the words is never completely innocent or arbitrary, and reflects the ideas that we have about the physical reality.

We have shown in this work that the photons in a medium are subjected to a force, proportional to the time derivative of the refractive index. Furthermore, their dynamical interaction with electrostatic waves in a plasma is very similar to that of a charged particle (an electron or an ion) interacting with the same wave.

In particular, photons can oscillate and can be trapped in the wavefield. On the other hand, the plasma waves can be damped by photon Landau damping, in the same way as they are Landau damped by the electrons.

We have also found that an effective photon mass, and an equivalent electric charge (or a dipole) for the photons in a medium, could be defined. This shows that our familiar view of photons as particles with no rest mass, and with no electric charge, can only apply to 'bare' photons moving in a vacuum and not to 'dressed' photons moving in a background medium. This means that we should not deny for photons what we accept as true for other particles: that, by receiving energy from the fields with which they interact, they are energized or, in other words, they are accelerated.

We can still argue that photons correspond to a particle description of the electromagnetic field, and that more generally, a wave description is necessary. But the same is also true for the other fields and for the other particles.

This means that the use of photon acceleration as a genuine physical concept can lead to a more global view of the physical processes and of the elementary interactions between the various particles and the various fields. What, at first sight, is seen as a more fashionable choice of terminology can lead us to ask new questions and eventually to get a deeper understanding about physics.

As an example, we could say that the equivalent charge of a photon in a plasma is nothing but a different way of describing the well-known ponderomotive force, or radiation pressure effects. However, the fact that we were able to isolate the new concept of an equivalent charge led us immediately to the problem of secondary radiation emitted by accelerated photons, such as the photon ondulator effects of the photon transition radiation. Other questions related to this concept, but not considered here, are the possibility of photon bending in a static magnetic field or the attraction between two parallel photon beams.

We can also explore the idea of photon acceleration as a particular example of particle acceleration by a time-varying mean field. Such a mean field process could operate not only with photons and the electromagnetic field, but also with other particles and with other fields. We saw in this chapter that the ideas of photon acceleration in an optical medium can be extrapolated to the case of photons interacting in a vacuum with a gravitational field, or to neutrinos moving in a dense plasma.

These different mean field processes involve the electromagnetic, the weak and the gravitational interactions. Similar processes can also be found for the strong interaction. In particular, the possibility of particle acceleration by the non-stationary nuclear matter produced by relativistic heavy ion collisions [20] is presently being explored [96].

Appendix

Derivation of the Wigner–Moyal equation

A.1 Non-dispersive media

We consider transverse electromagnetic fields ($\nabla \cdot \vec{E} = 0$), in a non-dispersive medium, as described by

$$\nabla^2 \vec{E} - \frac{1}{c^2} \frac{\partial^2 \vec{E}}{\partial t^2} = \frac{1}{c^2} \frac{\partial^2}{\partial t^2} (\chi \vec{E}). \tag{A.1}$$

Now, we use the notation $\vec{E}_i \equiv \vec{E}(\vec{r}_i, t_i)$ and $\chi_i \equiv \chi(\vec{r}_i, t_i)$, for $i = 1, 2$, and we write

$$\left(\nabla_1^2 - \frac{1}{c^2} \frac{\partial^2}{\partial t_1^2} \right) \vec{E}_1 = \frac{1}{c^2} \frac{\partial^2}{\partial t_1^2} \chi_1 \vec{E}_1 \tag{A.2}$$

$$\left(\nabla_2^2 - \frac{1}{c^2} \frac{\partial^2}{\partial t_2^2} \right) \vec{E}_2 = \frac{1}{c^2} \frac{\partial^2}{\partial t_2^2} \chi_2 \vec{E}_2. \tag{A.3}$$

Let us multiply the first of these equations by $\vec{E}_2^*$ and the complex conjugate of the second one by $\vec{E}_1$. Noticing that, in the absence of losses, the refractive index is always real and for this reason we can write $\chi_i = \chi_i^*$, we obtain, after subtracting the resulting two equations,

$$\left[(\nabla_1^2 - \nabla_2^2) - \frac{1}{c^2} \left(\frac{\partial^2}{\partial t_1^2} - \frac{\partial^2}{\partial t_2^2} \right) \right] C_{12} = \frac{1}{c^2} \left(\frac{\partial^2}{\partial t_1^2} \chi_1 - \frac{\partial^2}{\partial t_2^2} \chi_2 \right) C_{12} \tag{A.4}$$

with

$$C_{12} = \vec{E}_1 \cdot \vec{E}_2^*. \tag{A.5}$$

For convenience, let us introduce new space and time variables, such that

$$\vec{r} = \frac{1}{2}(\vec{r}_1 + \vec{r}_2), \quad \vec{s} = \vec{r}_1 - \vec{r}_2 \tag{A.6}$$

195

and

$$t = \frac{1}{2}(t_1 + t_2), \quad \tau = t_1 - t_2. \tag{A.7}$$

In an equivalent manner, we could have stated that

$$\vec{r}_1 = \vec{r} + \frac{\vec{s}}{2}, \quad \vec{r}_2 = \vec{r} - \frac{\vec{s}}{2} \tag{A.8}$$

and

$$t_1 = t + \frac{\tau}{2}, \quad t_2 = t - \frac{\tau}{2}. \tag{A.9}$$

Using these variable transformations, we can easily realize that

$$\left(\frac{\partial^2}{\partial t_1^2} \chi_1 - \frac{\partial^2}{\partial t_2^2} \chi_2 \right) = \left(\frac{1}{4} \frac{\partial^2}{\partial t^2} + \frac{\partial^2}{\partial \tau^2} \right) (\chi_1 - \chi_2) + \frac{\partial^2}{\partial t \partial \tau} (\chi_1 + \chi_2). \tag{A.10}$$

This expression can be simplified by noting that τ is a fast timescale and t is a slow timescale, as will become more obvious in the following. Furthermore, we can assume that the susceptibility χ is a slowly varying function and that its dependence on the fast time variable τ is negligible. Using $(\chi_1 + \chi_2) \simeq 2\chi$, we can then write this equation as

$$2 \left(\nabla \cdot \nabla_s - \frac{\epsilon}{c^2} \frac{\partial^2}{\partial t \partial \tau} \right) C_{12} \simeq \frac{1}{c^2} (\chi_1 - \chi_2) \frac{\partial^2}{\partial \tau^2} C_{12} + 2 \frac{\partial \chi}{\partial t} \frac{\partial}{\partial \tau} C_{12}. \tag{A.11}$$

We know that, by making a Taylor expansion of a function of time $f(t + \tau)$ around $f(t)$, we can obtain

$$f(t + \tau) = f(t) + \sum_{m=1}^{\infty} \frac{1}{m!} \tau^m \frac{\partial^m}{\partial t^m} f(t) \simeq f(t) + \tau \frac{\partial f(t)}{\partial t} + \cdots. \tag{A.12}$$

Introducing an exponential operator, this can be written in a more elegant and more compact form as

$$f(t + \tau) = \exp\left(\tau \frac{\partial}{\partial t} \right) f(t). \tag{A.13}$$

A power series development of this exponential operator clearly shows that this is equivalent to equation (A.12). Similarly, a function of the coordinates $f(\vec{r} + \vec{s})$ can be expanded around $f(\vec{r})$ as

$$f(\vec{r} + \vec{s}) = \exp(\vec{s} \cdot \nabla) f(\vec{r}). \tag{A.14}$$

This means that, by performing a double (space and time) Taylor expansion of the susceptibilites χ_1 and χ_2 around $\vec{r}$ and t, we obtain

$$\chi_1 = \chi(\vec{r} + \vec{s}/2, t + \tau/2) = \exp\left(\frac{\vec{s}}{2} \cdot \nabla + \frac{\tau}{2} \frac{\partial}{\partial t} \right) \chi(\vec{r}, t) \tag{A.15}$$

and

$$\chi_2 = \chi(\vec{r} - \vec{s}/2, t - \tau/2) = \exp\left(-\frac{\vec{s}}{2} \cdot \nabla - \frac{\tau}{2}\frac{\partial}{\partial t}\right)\chi(\vec{r}, t). \tag{A.16}$$

This means that the difference between the two values of the susceptibility of the medium can be written as

$$(\chi_1 - \chi_2) = 2\sinh\left(\frac{\vec{s}}{2} \cdot \nabla + \frac{\tau}{2}\frac{\partial}{\partial t}\right)\chi$$

$$= 2\sum_{l=0}^{\infty}\frac{1}{(2l+1)!}\left[\left(\frac{\vec{s}}{2} \cdot \nabla\right) + \left(\frac{\tau}{2}\frac{\partial}{\partial t}\right)\right]^{2l+1}\chi. \tag{A.17}$$

At this point it is useful to introduce the double Fourier transformation of C_{12}:

$$C_{12} \equiv C(\vec{r}, \vec{s}, t, \tau) = \int\frac{d\vec{k}}{(2\pi)^3}\int\frac{d\omega}{2\pi}F(\vec{r}, t; \omega, \vec{k})\,e^{i\vec{k}\cdot\vec{s}-i\omega\tau}. \tag{A.18}$$

The corresponding inverse transformation is related to the electric field as follows:

$$F(\vec{r}, t; \omega, \vec{k}) = \int d\vec{s}\int d\tau\,C(\vec{r}, \vec{s}, t, \tau)\,e^{-i\vec{k}\cdot\vec{s}+i\omega\tau}$$

$$= \int d\vec{s}\int d\tau\,\vec{E}\left(\vec{r} + \frac{\vec{s}}{2}, t + \frac{\tau}{2}\right) \cdot \vec{E}^*\left(\vec{r} - \frac{\vec{s}}{2}, t - \frac{\tau}{2}\right)e^{-i\vec{k}\cdot\vec{s}+i\omega\tau}. \tag{A.19}$$

This quantity is the Wigner function for the electric field. Using this definition in equation (A.11), we obtain

$$\left(\frac{\partial}{\partial t} + \frac{c^2\vec{k}}{\omega\epsilon} \cdot \nabla\right)F + \frac{\partial\ln\epsilon}{\partial t}F = i\frac{\omega}{2\epsilon}(\chi_1 - \chi_2)F. \tag{A.20}$$

Using equation (A.17), we can write on the right-hand side of this equation

$$(\chi_1 - \chi_2)F = 2\sum_{l=0}^{\infty}\frac{F}{(2l+1)!}\left[\left(\frac{\vec{s}}{2} \cdot \nabla\right) + \left(\frac{\tau}{2}\frac{\partial}{\partial t}\right)\right]^{2l+1}\chi. \tag{A.21}$$

But, from the definition of F, we can also write

$$\frac{\partial^m}{\partial\vec{k}^m} = (-i\vec{s})^m F, \qquad \frac{\partial^m}{\partial\omega^m} = (i\tau)^m F. \tag{A.22}$$

This means that we can rewrite equation (A.21) as

$$(\chi_1 - \chi_2)F = 2\sum_{l=0}^{\infty}\frac{(-1)^l}{(2l+1)!}\left[\frac{1}{2}\frac{\partial}{\partial\vec{k}} \cdot \nabla = -\frac{1}{2}\frac{\partial}{\partial\omega}\frac{\partial}{\partial t}\right]^{2l+1}\chi F. \tag{A.23}$$

Using this result in equation (A.20) we finally obtain

$$\left(\epsilon \frac{\partial}{\partial t} + \frac{c^2 \vec{k}}{\omega} \cdot \nabla\right) F + \left(\frac{\partial \epsilon}{\partial t}\right) F = -\omega(\epsilon \sin \Lambda F) \qquad (A.24)$$

with

$$\Lambda = \frac{1}{2} \overleftarrow{\left[\frac{\partial}{\partial \vec{r}} \cdot \frac{\partial}{\partial \vec{k}} - \frac{\partial}{\partial t} \frac{\partial}{\partial \omega}\right]} \overrightarrow{} . \qquad (A.25)$$

For a linear wave spectrum, we have

$$F \equiv F(\vec{r}, t; \omega, \vec{k}) = F_k(\vec{r}, t)\delta(\omega - \omega_k). \qquad (A.26)$$

Using this in the definition of C_{12}, we get

$$C_{12} \equiv C(\vec{r}, \vec{s}, t, \tau) = e^{-i\omega_k \tau} \int F_k(\vec{r}, t) e^{i\vec{k}\cdot\vec{s}} \frac{d\vec{k}}{(2\pi)^3}$$
$$= e^{-i\omega_k \tau} C(\vec{r}, \vec{s}, t, \tau = 0). \qquad (A.27)$$

According to equation (A.19), this means that we can define $F_k(\vec{r}, t)$ as the space Wigner function for the electric field:

$$F_k(\vec{r}, t) = \int C(\vec{r}, \vec{s}, t, \tau = 0) e^{-i\vec{k}\cdot\vec{s}} \, d\vec{s}$$
$$= \int \vec{E}(\vec{r} + \vec{s}/2, t) \cdot \vec{E}^*(\vec{r} - \vec{s}/2, t) e^{-i\vec{k}\cdot\vec{s}} \, d\vec{s}. \qquad (A.28)$$

A.2 Dispersive media

In a dispersive medium, we have

$$\left(\nabla^2 - \frac{1}{c^2}\frac{\partial^2}{\partial t^2}\right) \vec{E} = \mu_0 \frac{\partial^2}{\partial t^2} \vec{P} \qquad (A.29)$$

with $\vec{P} = \epsilon_0 \vec{E} - \vec{D}$.

Returning to the procedure followed in appendix A.1, we can see that equations (A.2, A.3) have to be replaced by

$$\left(\nabla_i^2 - \frac{1}{c^2}\frac{\partial^2}{\partial t_i^2}\right) \vec{E}_i = \mu_0 \frac{\partial^2}{\partial t_i^2} \vec{P}_i \qquad (A.30)$$

for $i = 1, 2$.

Again, we can derive from here an evolution equation for the quantity $C_{12} = \vec{E}_1 \cdot \vec{E}_2^*$. The result is

$$\left[(\nabla_1^2 - \nabla_2^2) - \frac{1}{c^2} \left(\frac{\partial^2}{\partial t_1^2} - \frac{\partial^2}{\partial t_2^2} \right) \right] C_{12}$$

$$= \mu_0 \left[\frac{\partial^2}{\partial t_1^2} (\vec{P}_1 \cdot \vec{E}_2^*) - \frac{\partial^2}{\partial t_2^2} (\vec{P}_2^* \cdot \vec{E}_1) \right]. \tag{A.31}$$

Let us now introduce the space and time variables defined by equations (A.6, A.7). This equation becomes

$$2 \left(\nabla \cdot \nabla_s - \frac{1}{c^2} \frac{\partial^2}{\partial t \partial \tau} \right) C_{12}$$

$$= \mu_0 \left(\frac{1}{4} \frac{\partial^2}{\partial t^2} + \frac{\partial^2}{\partial \tau^2} \right) \left(\vec{P}_1 \cdot \vec{E}_2^* - \vec{P}_2^* \cdot \vec{E}_1 \right)$$

$$+ \mu_0 \frac{\partial^2}{\partial t \partial \tau} \left(\vec{P}_1 \cdot \vec{E}_2^* + \vec{P}_2^* \cdot \vec{E}_1 \right). \tag{A.32}$$

Here we can introduce the Fourier transformation

$$\vec{E}_i \equiv \vec{E}(\vec{r}_i, t_i) = \int \frac{d\omega_i}{2\pi} \int \frac{d\vec{k}_i}{(2\pi)^3} \vec{E}(\omega_i, \vec{k}_i) \, e^{i\vec{k}_i \cdot \vec{r}_i - i\omega_i t_i}. \tag{A.33}$$

A similar transformation for the polarization vector is defined by

$$\vec{P}_i \equiv \vec{P}(\vec{r}_i, t_i) = \int \frac{d\omega_i}{2\pi} \int \frac{d\vec{k}_i}{(2\pi)^3} \vec{P}(\vec{r}_i, t_i; \omega_i, \vec{k}_i) \, e^{i\vec{k}_i \cdot \vec{r}_i - i\omega_i t_i} \tag{A.34}$$

such that

$$\vec{P}(\vec{r}_i, t_i; \omega_i, \vec{k}_i) = \epsilon_0 \chi(\vec{r}_i, t_i; \omega_i, \vec{k}_i) \vec{E}(\omega_i, \vec{k}_i). \tag{A.35}$$

The susceptibility of the medium $\chi(\vec{r}, t; \omega, \vec{k})$, appearing in this expression, is assumed to be a slowly varying function of space and time. We can rewrite the quantity C_{12} in terms of the Fourier components of the electric field. But because this would lead to quite cumbersome expressions, we prefer to introduce new frequency and wavevector variables, such that

$$\vec{q} = \vec{k}_1 + \vec{k}_2, \quad \vec{k} = \frac{1}{2}(\vec{k}_1 - \vec{k}_2) \tag{A.36}$$

and

$$\Omega = \omega_1 + \omega_2, \quad \omega = \frac{1}{2}(\omega_1 - \omega_2). \tag{A.37}$$

Equivalently, we could have stated that

$$\vec{k}_1 = \frac{\vec{q}}{2} + \vec{k}, \quad \vec{k}_2 = \frac{\vec{q}}{2} - \vec{k} \tag{A.38}$$

and

$$\omega_1 = \frac{\Omega}{2} + \omega, \quad \omega_2 = \frac{\Omega}{2} - \omega. \tag{A.39}$$

As in the case of the space and time variable transformations (A.6–A.9), the Jacobian of the new transformations is equal to one: $d\omega_1 \, d\omega_2 = d\Omega \, d\omega$ and $d\vec{k}_1 \, d\vec{k}_2 = d\vec{q} \, d\vec{k}$. In terms of these new variables, the quantity C_{12} becomes formally identical to equation (A.18), as it should be, with the quantity $F(\vec{r}, t; \omega, \vec{k})$ defined now as

$$F(\vec{r}, t; \omega, \vec{k}) = \int \frac{d\Omega}{2\pi} \int \frac{d\vec{q}}{(2\pi)^3} J(\vec{q}, \vec{k}, \Omega, \omega) \, e^{i\vec{q}\cdot\vec{r} - i\Omega t} \tag{A.40}$$

with

$$J(\vec{q}, \vec{k}, \Omega, \omega) = \vec{E}(\omega + \Omega/2, \vec{k} + \vec{q}/2) \cdot \vec{E}(-\omega + \Omega/2, -\vec{k} + \vec{q}/2). \tag{A.41}$$

Returning to equation (A.32) and retaining on its right-hand side only the dominant term, the one proportional to $\partial^2/\partial\tau^2$, we can write

$$2\left[\nabla \cdot \nabla_s - \frac{1}{c^2}\frac{\partial^2}{\partial t \partial \tau}\right] C_{12} = -\frac{1}{c^2} \int \frac{d\Omega}{2\pi} \int \frac{d\vec{q}}{(2\pi)^3} (\eta_+ - \eta_-)$$
$$\times J(\vec{q}, \vec{k}, \Omega, \omega) \, e^{i\vec{q}\cdot\vec{r} - i\Omega t} \, e^{i\vec{k}\cdot\vec{s} - i\omega\tau} \tag{A.42}$$

where, in order to simplify the expression, we have introduced the quantities

$$\eta_\pm = \left(\omega \pm \frac{\Omega}{2}\right)^2 \chi\left(\vec{r} \pm \frac{\vec{s}}{2}, t \pm \frac{\tau}{2}; \omega \pm \frac{\Omega}{2}, \vec{k} \pm \frac{\vec{q}}{2}\right). \tag{A.43}$$

Here, we should notice that $|\omega| \gg |\Omega|$ because ω is associated with the fast timescale τ, whereas the frequency Ω is associated with the slow timescale t. In the same way, we can assume that $|\vec{k}| \gg |\vec{q}|$. Developing these quantities around the values $(\omega, \vec{k})$ and $(\vec{r}, t)$, we obtain

$$\eta_\pm \simeq \eta_0 \pm \frac{\Omega}{2}\frac{\partial\eta_0}{\partial\omega} \pm \frac{\vec{q}}{2} \cdot \frac{\partial\eta_0}{\partial\vec{k}} \pm \frac{\tau}{2}\frac{\partial\eta_0}{\partial t} \pm \frac{\vec{s}}{2} \cdot \frac{\partial\eta_0}{\partial\vec{r}} + \cdots \tag{A.44}$$

where we have considered that

$$\eta_0 = \omega^2 \chi(\vec{r}, t; \omega, \vec{k}). \tag{A.45}$$

This means that, in equation (A.42), we can use

$$\eta_+ - \eta_- = \left(\Omega\frac{\partial\eta_0}{\partial\omega} + \vec{q} \cdot \frac{\partial\eta_0}{\partial\vec{k}}\right) + \left(\tau\frac{\partial\eta_0}{\partial t} + \vec{s} \cdot \frac{\partial\eta_0}{\partial\vec{r}}\right). \tag{A.46}$$

But we also notice that the quantity η_0 and its derivatives are independent of Ω and $\vec{q}$. It means that, in equation (A.42), they can be taken out of the integrations in these variables. This allows us to make the following replacements, in the same equation:

$$\int \frac{d\Omega}{2\pi} \int \frac{d\vec{q}}{(2\pi)^3} \Omega J(\vec{q},\vec{k},\Omega,\omega)\, e^{i\vec{q}\cdot\vec{r}-i\Omega t} = i\frac{\partial}{\partial t} F(\vec{r},t;\omega,\vec{k}) \qquad (A.47)$$

and

$$\int \frac{d\Omega}{2\pi} \int \frac{d\vec{q}}{(2\pi)^3} \vec{q}\, J(\vec{q},\vec{k},\Omega,\omega)\, e^{i\vec{q}\cdot\vec{r}-i\Omega t} = -i\nabla F(\vec{r},t;\omega,\vec{k}). \qquad (A.48)$$

The result is

$$2\left[\nabla\cdot\nabla_s - \frac{1}{c^2}\frac{\partial^2}{\partial t\,\partial\tau}\right]C_{12}$$
$$= -\frac{1}{c^2}\int \frac{d\omega}{2\pi}\int \frac{d\vec{k}}{(2\pi)^3}\left[i\left(\frac{\partial\eta_0}{\partial\omega}\frac{\partial}{\partial t}-\frac{\partial\eta_0}{\partial\vec{k}}\cdot\nabla\right)F(\vec{r},t;\omega,\vec{k})\right.$$
$$\left.+\left(\tau\frac{\partial\eta_0}{\partial t}+\vec{s}\cdot\nabla\eta_0\right)F(\vec{r},t;\omega,\vec{k})\right]e^{i\vec{k}\cdot\vec{s}-i\omega\tau}. \qquad (A.49)$$

We can now replace $C_{12} \equiv C(\vec{r},t,\vec{s},\tau)$ by its Fourier integral, as defined by equation (A.18), and we obtain

$$2i\left(\vec{k}\cdot\nabla+\frac{\omega}{c^2}\frac{\partial}{\partial t}\right)F(\vec{r},t;\omega,\vec{k}) = -\frac{i}{c^2}\left(\frac{\partial\eta_0}{\partial\omega}\frac{\partial}{\partial t}-\frac{\partial\eta_0}{\partial\vec{k}}\cdot\nabla\right)F(\vec{r},t;\omega,\vec{k})$$
$$-\frac{1}{c^2}\int d\vec{s}\int d\tau\int \frac{d\omega'}{2\pi}\int \frac{d\vec{k}'}{(2\pi)^3}\left(\tau\frac{\partial\eta_0}{\partial t}+\vec{s}\cdot\nabla\eta_0\right)$$
$$\times C(\vec{r},t,\vec{s},\tau)\, e^{i(\vec{k}'-\vec{k})\cdot\vec{s}}\, e^{-i(\omega'-\omega)\tau}. \qquad (A.50)$$

But, it is also clear that we can write

$$\int \tau\, e^{-i(\omega'-\omega)\tau}\, d\tau = i\frac{\partial}{\partial\omega'}\int e^{-i(\omega'-\omega)\tau}\, d\tau = 2\pi i\delta(\omega'-\omega)\frac{\partial}{\partial\omega'} \qquad (A.51)$$

and

$$\int \vec{s}\, e^{i(\vec{k}'-\vec{k}\cdot\vec{s})}\, d\vec{s} = -(2\pi)^3\, i\delta(\vec{k}'-\vec{k})\frac{\partial}{\partial\vec{k}'}. \qquad (A.52)$$

This means that we can finally transform equation (A.51) into a closed differential equation for the Wigner function $F \equiv F(\vec{r},t,\omega,\vec{k})$, which takes the form

$$2\omega\left(\frac{\partial}{\partial t}+\frac{c^2\vec{k}}{\omega}\cdot\nabla\right)F = -\left(\frac{\partial\eta_0}{\partial\omega}\frac{\partial F}{\partial t}-\frac{\partial\eta_0}{\partial\vec{k}}\cdot\nabla F\right)$$
$$+\left(\frac{\partial F}{\partial\omega}\frac{\partial\eta_0}{\partial t}-\frac{\partial F}{\partial\vec{k}}\cdot\nabla\eta_0\right). \qquad (A.53)$$

After rearranging the terms in this equation, we can rewrite it in a more suitable form:

$$\left(\frac{\partial}{\partial t} + \vec{v}_g \cdot \nabla\right) F = -\frac{2}{2\omega + \partial \eta_0 / \partial \omega} (\eta_0 \Lambda F) \qquad (A.54)$$

where Λ is defined by equation (A.25), and

$$\vec{v}_g = \frac{2c^2 \vec{k} - \omega^2 \partial \epsilon / \partial \vec{k}}{2\omega \epsilon + \omega^2 \partial \epsilon / \partial \omega} \qquad (A.55)$$

with $\epsilon = 1 + \chi = 1 + \eta_0 / \omega^2$.

References

[1] Abdullaev S S 1993 *Chaos and Dynamics of Rays in Waveguide Media* (London: Gordon and Breach)

[2] Abdullaev F 1994 *Theory of Solitons in Inhomogeneous Media* (New York: Wiley)

[3] Alfano R R (ed) 1989 *The Supercontinuum Laser Source* (New York: Springer)

[4] Ammosov M V, Delone N B and Krainov V P 1985 Tunnel ionization of complex atoms and of atomic ions in an alternating electromagnetic field *Sov. Phys.– JETP* **64** 1191

[5] Anderson P W 1963 Plasmons, gauge invariance, and mass *Phys. Rev.* **130** 439

[6] Bachelard G 1975 *La Formation de l'Esprit Scientifique* (Paris: Librairie Philosophique J Vrin)

[7] Banos A Jr, Mori W B and Dawson J M 1993 Computation of the electric and magnetic fields induced in a plasma created by ionization lasting a finite interval of time *IEEE Trans. Plasma Sci.* **21** 57

[8] Bernstein I B 1975 Geometric optics in space- and time-varying plasmas *Phys. Fluids* **18** 320

[9] Bernstein I B and Friedland L 1983 Geometric optics in space and time varying plasmas *Handbook of Plasma Physics* vol 1, ed M N Rosenbluth and R Z Sagdeev (Amsterdam: North-Holland)

[10] Bethe H 1986 Possible explanation of the solar-neutrino puzzle *Phys. Rev. Lett.* **56** 1305

[11] Binette L, Joguet B and Wang J C L 1998 Evidence of Fermi acceleration of Lyman α in the radio galaxy 1243+036 *Astrophys. J.* **505** 634

[12] Bingham R, DeAngelis U, Amin M R, Cairns R A and MaNamara B 1992 Relativistic Langmuir waves generated by ultrashort laser pulses *Plasma Phys. Control. Fusion* **34** 557

[13] Bingham R, Dawson J M, Su J J and Bethe H A 1994 Collective interactions between neutrinos and dense plasmas *Phys. Lett.* A **193** 279

[14] Bingham R, Mendonça J T and Dawson J M 1997 Photon Landau damping *Phys. Rev. Lett.* **78** 247

[15] Born M and Wolf E 1970 *Principles of Optics* (Elmsford, NY: Pergamon)

[16] Budden K G 1961 *Radio Waves in the Ionosphere* (Cambridge: Cambridge University Press)

[17] Chirikov B V 1979 Universal instability of many-dimensional oscillator sytems *Phys. Rep.* **52** 263

[18] Chesters D 1973 Dispersion of gravitational waves by a collisionless gas *Phys. Rev.* D **7** 2863

[19] Cirone M, Rzazewski K and Mostowski J 1997 Photon generation by time-dependent dielectric: a soluble model *Phys. Rev.* A **55** 62

[20] Csernai L P 1994 *Introduction to Relativistic Heavy Ion Collisions* (Chichester: Wiley)

[21] Dias J M, Stenz C, Lopes N, Badiche X, Blasco F, Dos Santos A, Silva L O, Mysyrowicz A, Antonetti A and Mendonça J T 1997 Experimental evidence of photon acceleration of ultrashort laser pulses in relativistic ionization fronts *Phys. Rev. Lett.* **78** 4773

[22] Dias J M, Silva L O and Mendonça J T 1998 Photon acceleration versus frequency-domain interferometry for laser wakefield diagnostics *Phys. Rev. ST—Accelerators Beams* **1** 031301

[23] Diels J C and Rudolph W 1996 *Ultrashort Laser Pulse Phenomena* (New York: Academic)

[24] Dysthe F B, Pecseli H L and Trulsen J 1983 Stochastic generation of continuous wave spectra *Phys. Rev. Lett.* **50** 353

[25] Elliott R S *An Introduction to Guided Waves and Microwave Circuits* (Englewood Cliffs, NJ: Prentice-Hall)

[26] Esarey E, Ting A and Sprangle P 1990 Frequency shifts induced in laser pulses by plasma waves *Phys. Rev.* A **42** 3526

[27] Esarey E, Joyce G and Sprangle P 1991 Frequency up-shifting of laser pulses by co-propagating ionization fronts *Phys. Rev.* A **44** 3908

[28] Faro A and Mendonça J T 1982 Interaction of electromagnetic waves with a moving perturbation in a stationary gas *J. Phys. D: Appl. Phys.* **16** 287

[29] Fermi E 1949 On the origin of the cosmic radiation *Phys. Rev.* **75** 1169

[30] Figueira G, Mendonça J T and Silva L O 1996 Discrete Fermi mapping for photon acceleration in a time varying plasma cavity *Transport, Chaos and Plasma Physics 2* (Singapore: World Scientific) p 237

[31] Fisher D L and Tajima T 1993 Superluminous laser pulse in an active medium *Phys. Rev. Lett.* **71** 4338

[32] Fukuda Y *et al* 1999 Measurement of the flux and zenith-angle distribution of upward through going muons by Super Kamiokande *Phys. Rev. Lett.* **82** 2644

[33] Goldstein H 1980 *Classical Mechanics* 2nd edn (Reading, MA: Addison-Wesley)

[34] Gorbunov L M and Kirsanov V I 1987 Excitation of plasma waves by an electromagnetic wave packet *Sov. Phys.– JETP* **66** 290

[35] Gradshteyn I S and Ryzhik I M 1980 *Table of Integrals, Series and Products* (New York: Academic) pp 307, 338

[36] Granatstein V L, Sprangle P, Parker R K, Pasour J, Herndon M, Schlesinger S P and Seftor J L 1976 Realization of a relativistic mirror: electromagnetic backscattering from the front of a magnetized relativistic electron beam *Phys. Rev.* A **14** 1194

[37] Haken H 1981 *Light, Vol. 1: Waves, Photons, Atoms* (Amsterdam: North-Holland)

[38] Harris S E 1997 Electromagnetically induced transparency *Phys. Today* **50** 36

[39] Hau L V, Harris S E, Dutton Z and Behroozi C H 1999 Light speed reduction to 17 metres per second in an ultracold atomic gas *Nature* **397** 594

[40] Hillary M, O'Connel R F, Scully M O and Wigner E P 1984 Distribution functions in physics: fundamentals *Phys. Rep.* **106** 121

[41] Hizanidis K, Vomvoridis J L, Mendonça J T and Franszeskakis D J 1996 Relativistic theory of frequency blue-shift of an intense ionizing laser beam in a plasma *IEEE Trans. Plasma Sci.* **24** 323

[42] Special issue on Second Generation Plasma Accelerators 1996 *IEEE Trans. Plasma Sci.* **24** (2)

[43] Jackson J D 1999 *Classical Electrodynamics* 3rd edn (New York: Wiley)

[44] Kalluri D K 1999 *Electromagnetics of Complex Media: Frequency Shifting by a Transient Magnetoplasma Medium* (Boca Raton, FL: CRC)

[45] Kane D J and Trebino R 1993 Single-shot measurements of the intensity and phase of an arbitrary ultrashort pulse by using frequency resolved optical grating *Opt. Lett.* **18** 823

[46] Kaw P K, Sen A and Katsouleas T 1992 Nonlinear 1D laser pulse solitons in a plasma *Phys. Rev. Lett.* **68** 3172

[47] Keldysh L V 1965 Ionization in the field of a strong electromagnetic wave *Sov. Phys.–JETP* **20** 1307

[48] Kim A V, Lirin S F, Sergeev A M, Vanin E V and Stenflo L 1990 Compression and frequency up-conversion of an ultrashort ionizing pulse in a plasma *Phys. Rev. A* **42** 2493

[49] Klimontovich Yu L 1967 *The Statistical Theory of Non-equilibrium Processes in a Plasma* (Cambridge, MA: MIT Press)

[50] Kruer W L 1988 *The Physics of Laser Plasma Interactions* (Redwood City, CA: Addison-Wesley)

[51] Kuhn T S 1970 *The Structure of Scientific Revolutions* 2nd edn (Chicago, IL: University of Chicago Press)

[52] Kuo S P and Ren A 1993 Experimental study of wave propagation through a rapidly created plasma *IEEE Trans. Plasma Sci.* **21** 53

[53] Lampe M, Ott E and Walker J H 1978 Interaction of electromagnetic waves with a moving ionization front *Phys. Fluids* **21** 42

[54] Landau L D 1946 On the vibration of the electric plasma *J. Phys. (USSR)* **10** 25

[55] Landau L D and Lifshitz E M 1996 *The Classical Theory of Fields* (Oxford: Butterworth-Heinemann)

[56] Landau L D and Lifshitz E M 1994 *Mechanics* 3rd edn (Oxford: Butterworth-Heinemann)

[57] Landau L D and Lifsjitz E M 1984 *Electrodynamics of Continuous Media* 2nd edn (Oxford: Butterworth-Heinemann)

[58] Leonhardt U 1997 *Measuring the Quantum State of Light* (Cambridge: Cambridge University Press)

[59] Lichtenberg A J and Lieberman M A 1983 *Regular and Stochastic Motion* (New York: Springer)

[60] Lotz W 1967 An empirical formula for the electron impact ionization cross-section *Z. Phys.* **206** 205

[61] Loudon R 1983 *The Quantum Theory of Light* (Oxford: Clarendon)

[62] Mendonça J T 1979 Nonlinear transition radiation in a plasma *Phys. Rev. Lett.* **43** 354

[63] Mendonça J T 1979 Nonlinear interaction of wavepackets *J. Plasma Phys.* **22** 15

[64] Mendonça J T 1985 Beat-wave excitation of plasma waves *J. Plasma Phys.* **34** 115

[65] Mendonça J T, Hisanidis K, Frantzeskakis D J, Silva L O and Vomvoridis Y L 1997 Covariant formulation of photon acceleration *J. Plasma Phys.* **58** 647

[66] Mendonça J T and Silva L O 1994 Regular and stochastic acceleration of photons *Phys. Rev. E* **49** 3520

[67] Mendonça J T and Silva L O 1995 Mode coupling theory of flash ionization *IEEE Trans. Plasma Sci.*

[68] Mendonça J T, Bingham R, Shukla P K, Dawson J M and Tsytovich V N 1995 Interaction between neutrinos and nonstationary plasmas *Phys. Lett.* A **209** 78

[69] Mendonça J T, Silva L O, Bingham R, Tsintsazde N L, Shukla P K and Dawson J M 1998 Effective charge of photons and neutrinos in a plasma *Phys. Lett.* A **239** 373

[70] Mendonça J T, Hizanidis K and Franszeskakis D J 1998 Method for generating tunable high-frequency harmonics in periodically modulated $\xi^{(2)}$ materials *Opt. Commun.* **146** 245

[71] Mendonça J T, Shukla P K, Bingham R and Tsintsadze N L 1999 Transition radiation of photons and neutrinos at a plasma boundary *Phys. Scr.* at press

[72] Mendonça J T, Shukla P K and Bingham R 1998 Photon acceleration by gravitational waves *Phys. Lett.* A **250** 144

[73] Mendonça J T, Shukla P K, Bingham R, Dawson J M and Silva L O 1999 Neutrino Landau damping and collective neutrino–plasma processes *J. Plasma Phys.* at press

[74] Mendonça J T, Guerreiro A and Martins A M 1999 Quantum theory of time refraction, in preparation

[75] Meszaros P, Laguna P and Rees M J 1993 Gas dynamics of relativistically expanding gamma-ray burst sources: kinematics, energetics, magnetic fields, and efficiency *Astrophys. J.* **415** 181

[76] Misner C W, Thorne K S and Wheeler J A 1970 *Gravitation* (San Francisco: Freeman)

[77] Mori W B 1987 On beat-wave excitation of relativistic plasma waves *IEEE Trans. Plasma Sci.* **15** 88

[78] Mori W B 1991 Generation of tunable radiation using an underdense ionization front *Phys. Rev. Lett.* **44** 5118

[79] Mori W B, Katsouleas T, Dawson J M and Lai C H 1995 Conversion of dc fields in a capacitor array to radiation by a relativistic ionization front *Phys. Rev. Lett.* **74** 542

[80] Moyal J E 1949 Quantum mechanics as a statistical theory *Proc. Camb. Phil. Soc.* **45** 99

[81] Nayfeh A 1993 *Introduction to Perturbation Theory* (New York: Wiley) p 253

[82] Nicholson D R 1983 *Introduction to Plasma Physics* (New York: Wiley)

[83] Nieves J F and Pal P B 1994 Induced charge of neutrinos in a medium *Phys. Rev.* D **49** 1398

[84] Olver F W J 1965 *Handbook of Mathematical Functions* ed M Abramowitz and I A Stegun (New York: Dover) p 355

[85] Oraevskii V N, Semikov V B and Smorodinskii Ya A 1994 Electrodynamics of neutrinos in a medium *Phys. Part. Nucl.* **25** 1063

[86] Pais A 1982 *'Subtle is the Lord...' the Science and Life of Albert Einstein* (Oxford: Oxford University Press)

[87] Pasour J A, Granatstein V L and Parker R K 1997 'Relativistic mirror' experiment with frequency tuning and energy gain *Phys. Rev.* A **16** 2441

[88] Quang Ho-Kim and Pham Xuan Yem 1998 *Elementary Particles and Their Interactions* (Berlin: Springer)

[89] Rax J M and Fisch N J 1993 Ultrahigh intensity laser–plasma interaction: a Lagrangian approach *Phys. Fluids* B **5** 2578

[90] Rivlin L A 1997 Photons in a waveguide (some thought experiments) *Phys. Usp.* **40** 291

[91] Rosa C C, Oliveira e Silva L, Lopes N and Mendonça J T 1998 Formation of ionization fronts by intense short laser pulses propagating in a neutral gas *Superstrong Fields in Plasmas* ed M Lontano, G Mourou, F Pegoraro and E Sindoni (Verena: American Institute of Physics) pp 135–40

[92] Rosenbluth M N and Liu C S 1972 Excitation of plasma waves by two laser beams *Phys. Rev. Lett.* **29** 701

[93] Sagdeev R Z and Galeev A A 1969 *Nonlinear Plasma Theory* (New York: Benjamin)

[94] Santos D J and Loudon R 1995 Electromagnetic field quantization in inhomogeneous and dispersive one-dimensional systems *Phys. Rev.* A **52** 1538

[95] Savage R L Jr, Joshi C and Mori W B 1992 Frequency upconversion of electromagnetic waves upon transmission into an ionization front *Phys. Rev. Lett.* **68** 946

[96] Seixas J and Mendonça J T 1999 Moving boundary effects in QCD plasma formation, in preparation

[97] Semenova V I 1967 Reflection of electromagnetic waves from an ionization front *Sov. Radiophys. Quantum Electron.* **10** 599

[98] Silva L O and Mendonça J T 1992 Asymmetric pendulum *Phys. Rev.* A **46** 6700

[99] Silva L O and Mendonça J T 1996 Photon acceleration in superluminous and accelerated ionization fronts *IEEE Trans. Plasma Sci.* **24** 316

[100] Silva L O and Mendonça J J 1996 Full wave theory of photon acceleration in a cavity *IEEE Trans. Plasma Sci.* **24** 503

[101] Silva L O, Mendonça J T and Figueira G 1996 Full wave theory of Fermi photon acceleration *Phys. Scr.* T **63** 288

[102] Silva L O, Bingham R, Dawson J M, Mendonça J T and Shukla P K 1999 Neutrino driven streaming instabilities in a dense plasma *Phys. Rev. Lett.* **83** 2703

[103] Silva L O and Mendonça J T 1999 Where is phase in self-phase modulation? to be submitted

[104] Shukla P K, Tsintsadze N L, Mendonça J T and Stenflo L 1999 Equivalent charge of photons in magnetized plasmas *Phys. Plasmas* **6** 627

[105] Shukla P K, Bingham R, Mendonça J T and Stenflo L 1998 Neutrinos generating inhomogeneities and magnetic fields in the early universe *Phys. Plasmas* **5** 2815

[106] Shukla P K, Silva L O, Bethe H, Bingham R, Dawson J M, Stenflo L, Mendonça J T and Dalhed S 1999 The physics of collective neutrino–plasma interactions *Plasma Phys. Control. Fusion* **41** A699

[107] Solokov A A, Mora P and Chessa P 1999 Simulation of photon acceleration in a plasma wakefield *Phys. Plasmas* **6** 503

[108] Stix T H 1962 *The Theory of Plasma Waves* (New York: McGraw-Hill)

[109] Tajima T and Dawson J M 1979 Laser electron accelerator *Phys. Rev. Lett.* **43** 267

[110] Tsai C S and Ault B A 1967 Wave interactions with moving boundaries *J. Appl. Phys.* **38** 2106

[111] Tsintadze N L and Mendonça J T 1998 Kinetic theory of photons in a plasma *Phys. Plasmas* **5** 3609

[112] Tsintsadze N L, Mendonça J T and Tsintsadze L N 1998 Kinetic formulation of neutrino–plasma interactions *Phys. Plasmas* **5** 3512

[113] Tsintsadze N L, Mendonça J T and Shukla P K 1998 Kinetic theory of induced electric charges of intense photons and neutrinos in an unmagnetized plasma *Phys. Lett.* A **249** 110

[114] Tsytovich V N 1970 *Nonlinear Effects in Plasmas* (New York: Plenum)

[115] van Ojik R, Roettgering H J A, Miley G K and Hunstead R W 1997 The gaseous environments of radio galaxies in the early universe: kinematics of the Lyman α emission and spatially resolved HI absorption *Astron. Astrophys.* **317** 358

[116] Walls D F and Milburn G J 1995 *Quantum Optics* (Berlin: Springer)

[117] Weinberg S 1962 Eikonal method in magnetohydrodynamics *Phys. Rev.* **126** 1899

[118] Wilks S C, Dawson J M, Mori W B, Katsouleas T and Jones M E 1989 Photon accelerator *Phys. Rev. Lett.* **62** 2600

[119] Wolfenstein L 1978 Neutrino oscillations in matter *Phys. Rev.* D **17** 1369

[120] Wood W M, Siders C W and Downer M C 1993 Femtosecond growth dynamics of an underdense ionization front measured by spectral blueshifting *IEEE Trans. Plasma Sci.* **21** 20

[121] Xu X, Yugami N and Nishida Y 1997 Frequency upshift in the interaction of a high power microwave with an inhomogeneous plasma *Phys. Rev.* A **55** 3328

[122] Yablonovitch E 1974 Self-phase modulation and short-pulse generation from laser-breakdown plasma *Phys. Rev.* A **10** 1888

[123] Xiong C, Yang Z and Liu S 1995 Frequency shifts by beam-driven plasma wake waves *Phys. Lett.* A **43** 203

[124] Zaslavsky G M 1988 *Chaos in Dynamical Systems* (New York: Harwood)

[125] Zaslavsky G M, Sagdeev R Z, Usikov D A and Chernikov A A 1991 *Weak Chaos and Quasi-Regular Patterns* (Cambridge: Cambridge University Press)

[126] Zeldovich Ya B and Novikov I D 1971 *Relativistic Astrophisics, Vol 1: Stars and Relativity* (Chicago, IL: University of Chicago Press)

Glossary

Dark source

This is the name of an experimental configuration where an ionization front radiates high-frequency electromagnetic waves by moving across a region filled with a static and spatially periodic electric field. This field can be produced by an array of capacitors with alternate polarities. In the language of photon acceleration, this radiation process can be seen as an acceleration of photons initially having zero frequency.

Flash ionization

When an intense laser pulse is focused into a region filled with a neutral gas it photoionizes the gas around the focal region and creates a plasma. This process of nearly instantaneous plasma creation, with the corresponding frequency up-shift of the laser frequency, is usually referred to as flash ionization. It should be contrasted with the ionization front case, where the boundaries of the plasma region can move with relativistic velocities. In some sense, flash ionization is a limiting case of an ionization front where the velocity of the front tends to infinity, because the whole plasma is created in the focal region simultaneously.

Frequency up-shift

Frequency up-shift (or the equally possible, but usually less exciting, frequency down-shift) is another name for photon acceleration. It comes naturally from the full-wave description of wave propagation in non-stationary media, where the concept of the photon is not explicitly considered. However, photon acceleration is a more suggestive concept, if we use geometric optics or quantum theory, or if we compare it with similar effects in other physical domains.

Induced phase modulation

This is the frequency shift of a probe laser pulse which propagates in a nonlinear optical medium in the presence of a second laser pulse. If the intensity of this second pulse is non-negligible, it produces a time-dependent phase modulation of the first pulse. If the two pulses have similar intensities they can mutually modulate their phases, and both shift their frequencies, thus leading to cross-phase modulation. These two processes of induced and cross-phase modulation can be seen as particular examples of photon acceleration.

Ionization front

This is the boundary between two distinct regions of a gaseous medium. On one side of the boundary the gas is neutral, and on the other side, the gas is ionized (a plasma). If the plasma state is created by an intense laser pulse propagating across a neutral gas, such a boundary can move with relativistic velocities, even if the particles in these two media stay nearly at rest. The interest of the ionization front is that it creates a moving discontinuity of the refractive index, leading to the acceleration (or frequency shift) of photons which interact with this moving boundary.

Landau damping

In a famous paper, Landau was able to prove that an electron plasma wave can be damped in a collisionless plasma. This damping process is due to the resonant interaction of the wave with the plasma electrons travelling with a velocity nearly equal to the wave phase velocity. Of course, for electron plasma waves with relativistic phase velocities (such as those associated with laser wakefields) the electron Landau damping is negligible because the number of resonant electrons tends to zero. However, in this case, the electron plasma waves can still be Landau damped, not by the electrons, but by the resonant photons. This surprising result is a consequence of the acceleration of those many photons resonantly interacting with a relativistic plasma wave. Its existence reveals the possibility of energy exchange between the non-stationary medium (in this case a plasma perturbed by the electron plasma wave) and the photon gas.

Magnetic mode

When an incident electromagnetic wave interacts with a moving ionization front it excites, not only the expected reflected and transmitted waves, but also a new kind of wave which is (in the laboratory frame of reference) a purely magnetic perturbation with zero frequency. This is called the magnetic mode. This mode has not yet been observed in experiments, but the theoretical arguments clearly point to its existence. In the language of photon acceleration we could say that the creation of a magnetic mode corresponds to a deceleration of the incident high-frequency photons down to zero-frequency photons.

Photon

The word photon has two different meanings, depending on whether we are referring to classical or quantum theory. In classical theory, it refers to an electromagnetic wavepacket with a length and time duration much shorter than the characteristic space and timescales of the medium. In the frame of geometric optics, the photon trajectories are determined by the ray equations. In quantum theory, it refers to the elementary excitation of the electromagnetic field, and can be identified with a quantum of electromagnetic energy and momentum.

Photon acceleration

This is the total energy variation of a photon when it moves in a non-stationary medium. The change in energy is equivalent to a shift in the photon frequency, with a change in the (group) velocity, which justifies the name. It should be noticed that the total energy of a photon can be divided in two parts: a kinetic part, associated with its wavevector, and a rest-mass part, associated with the polarization of the medium. In inhomogeneous but stationary media, the kinetic energy of the photon is not conserved, even if its total energy stays constant. In this case, we cannot talk about photon acceleration. Such an acceleration can only occur in non-stationary media.

Photon effective mass

A photon always moves in a material medium with a velocity less than the velocity of light in vacuum c. For that reason, we can say that the photon behaves like a particle with an effective mass. If the medium is an isotropic plasma, the photon mass is simply proportional to the electron plasma frequency. In general, it will be determined by the susceptibility of the medium. We should notice that, when we say *a photon in a medium* we are refering to a complex entity, which we can call a *dressed photon*, in contrast with the *bare photon* which only exists in a vacuum.

Photon equivalent charge

The electrons in a plasma are pushed away from the regions of intense electromagnetic energy (or regions containing a large photon density) by the ponderomotive force or radiation pressure. Alternatively, we can say that the photons behave as if they had a negative electric charge, thus pushing away the other negative charges (the electrons). This property can only be described with nonlinear equations, in contrast with the photon effective mass, which is intrinsically linear. We use here a different adjective (equivalent, instead of effective) in order to keep in mind the different nature of the photon charge and mass, one being linear and the other nonlinear. We can also derive a similar concept for a non-ionized medium. For instance, we can associate an electric dipole with the photons moving along an optical fibre. The main difference with respect to the plasma case is the non-existence of free electrons in the fibre.

Self-blueshift

The process of ionization of a neutral gas by an incident laser pulse produces a temporal change in the refractive index leading to an upshift, or a blueshift, of the laser frequency. This is a particular case of photon acceleration where the acceleration process is due to the incident photon beam itself.

Self-phase modulation

When an intense and short laser pulse propagates in a nonlinear optical medium (for instance along an optical fibre) it produces a nonlinear modulation of its own phase, thus leading to a significant frequency shift. The resulting spectral width can be significantly larger than that of the initial laser pulse and the final light pulses are sometimes referred to as supercontinuum radiation. Self-phase modulation can also be seen as a photon acceleration process.

Time reflection

A sudden change in the refractive index of a dielectric medium leads to the appearance of a wave with the same wavenumber as the pre-existing wave but propagating in the opposite direction. Fresnel formulae for time reflection, similar to the well-known Fresnel formulae for the usual (space) reflection, can be found.

Time refraction

A sudden change in the refractive index of a dielectric medium leads to a change in the frequency of the photons moving in this medium, but maintaining their wavevector. This is similar, but in some sense symmetric, to the case of the usual (space) refraction, where the refractive index changes in space and not in time, leading to the conservation of the photon frequency but a variation of the wavevector. A very simple and general Snell's law for time refraction can be found.

Wakefield

We call the laser wakefield, or more simply the wakefield, the fast electron plasma wave created by a short and intense laser pulse propagating in a plasma. The phase velocity of this electron plasma wave is nearly equal to the group velocity of the laser pulse.

Index

accelerated front, 39–42
accelerators, 1, 44
adiabatic process, 4
Alfven waves, 63
angle of,
    incidence, 19
    temporal incidence, 14
    transmission, 19
anti-Stokes sidebands, 95
area preserving map, 61
asymmetric,
    pendulum, 46
    states, 170
astrophysical,
    objects, 1, 186
    problems, 2, 63
atomic number, 83
atomic transitions,
    detuning, 65
averaged frequency, 93

'bare' particle, 97, 194, 211
beat-wave, 43
Bessel functions, 154
binomial expansion, 170
blueshift, 63, 67
Boltzmann distribution, 186
bounce frequency, 48
boundary,
    conditions, 174
    layer, 12
    moving, 175
bremsstrahlung, 3, 106, 107
broadband radiation, 4, 87

canonical,
    equations, 188, 192
    transformations, 15, 18, 39, 92
canonical variables, 10, 52
    covariant form, 29
capacitors, 136, 137, 209
    elongated, 65
Cartesian coordinates, 184
Cauchi principal part, 111
cavity, 57, 140, 152
centro-symmetric media, 118
Cherenkov radiation, 3
Chirikov criterion, 53
chirp, 67, 78, 90, 93, 94, 95
classical theory of radiation, 69
coarse-graining, 69
coherent states, 171, 172
commutation relations, 159, 160
completely ionized atoms, 83
conservation,
    equation, 67
    rules, 115
constants of motion, 15, 16, 18
continuity,
    conditions, 166
    equation, 84
co-propagating photons, 36, 40
cosmical scale, 190, 193
cosmic ray acceleration, 57
Coulomb,
    field, 187
    gauge, 158
coupling coefficients, 152, 182
counter-propagating photons, 36, 40

215

covariant formulation, 6, 49, 54
critical angle, 21
curved space–time, 29
cut-off,
    condition, 57, 128
    frequency, 38, 191
cyclotron resonance, 64

D'Alembert transformation, 179
dark source, 7, 132, 136, 209
density profile, 57
    parabolic, 60
deeply trapped trajectories, 47
Debye length, 43
dielectric,
    constant, 9, 157, 161, 165, 173,
        184
    function, 89, 120
    medium, 31, 115, 123, 157, 183
diffraction, 72, 74
diffusion, 32
Dirac equation, 179
discrete map, 57, 59
dispersion relation, 9, 43, 50, 52, 54,
        81, 130, 135, 164, 184, 186,
        191, 192
    general form, 69
    linear, 71
    of electron plasma waves, 109,
        113
dispersive medium, 9
Doppler shift, 43, 58
'dressed' particle, 97, 194, 211
dynamical systems, 53

early universe, 177, 183
effective potential, 179
eigen
    frequencies, 152
    functions, 150
    modes, 140, 150, 152, 167
eikonal equation, 23, 184
Einstein's
    energy formula, 25
    equations, 188, 191
    summation rule, 27
    theory of relativity, 19
electromagnetic,
    cavity, 57
    wavepacket, 97
    induced transparency, 65
electron,
    acceleration, 44
    beam, 149
    charge, 9
    cyclotron frequency, 63
    density, 9
    fluid equations, 98, 108, 162
    impact ionization, 83
    mass, 9
    plasma frequency, 9, 24, 33, 39,
        128, 178
    plasma wave, 6, 43, 97, 108, 112,
        126, 152, 164, 183, 193,
        211, 210, 213
    temperature, 43, 162, 186
    thermal velocity, 43, 108, 162,
        165
electron-impact ionization, 32
electrostatic waves, 43
energy,
    cascade, 149
    conservation law, 86–87
    density, 69
    eigenvalues, 167
envelope equation, 89
extraordinary mode, 64

Faraday equation, 98
Fermi,
    acceleration, 6, 57, 152–156
    constant, 178, 182
    mapping, 61
field,
    configuration, 9
    operators, 7
    phase, 127
    quantization, 157–165

fixed points, 45–47, 52
  stability, 60, 61
flash ionization, 126, 143, 149, **209**
flat space–time, 27
Fock states, 167–169
force acting on the photons, 109, 193
forced linear oscillator, 100
form function, 37
Fourier transformation, 119, 134, 180, 197, 199
four-vectors, 27, 28, 29, 49, 163, 184
frequency,
  cut-off, 9, 25
  local, 8
frequency shift, up-shift, 4, 39, 40, 58, 64, 146, 171, 172, 181, 190, 193, 209, **209**
  maximum value, 34, 47, 94
Fresnel formulae, 7, 123, 124, 212
  generalized, 126–128
  for time reflection, 126
  quantum, 174, 175
front,
  acceleration, 41
  profiles, 33
  velocity, 40
  width, 33
full wave models, 5, 209

gamma rays, 189, 192
Gaussian,
  distribution, 77, 78
  pulse, 33, 83, 90, 94, 106
geometric optics, 5, 69, 71, 123, 178, 209
generating function, 16, 18, 39
glass, 63
gravitational,
  field, 7, 29, 177, 183, 194
  lens, 186, 192
  redshift, 183–186, 192
  shock, 189, 192

wavepackets, 186, 187, 193
group velocity, 8, 10, 13, 23, 25, 61, 74, 77, 81, 90, 105, 112, 113, 213
growth rates, 114

Hamiltonian, 6, 10, 11, 23, 33, 40, 47, 50, 52, 64, 92, 156, 185, 186, 187, 188, 191
  covariant, 49
  operator, 159, 160
  theory, 183
harmonic generation, 100, 117, 122
Heaviside function, 12, 133, 147, 152, 166
heavy ion collisions, 194
helicity states, 179
Higgs boson, 165
hyperbolic tangent model, 12

ideal gas, 85
inhomogeneous medium, 11
instabilities,
  decay, 115
  dark source radiation, 139
  photon beam-plasma, 113–115
  wave, 112
intensity profile, 66
interactions,
  elementary, 194
  gravitational, 3
  laser–plasma, 6
  laser–matter, 1
  neutrino–plasma, 7, 177
  strong, 3, 194
  weak, 3, 177, 178
interaction time, 42
intermittency, 57
invariants, 28, 46, 47, 39, 50, 178, 181, 185, 189, 190
ionization,
  flash, 5, 36
  front, 4, 31-39, 67, 82–84, 128, 133, 138, 146, 148, 141, 209, **210**

probabilities, 82, 83
single step, 33, 38

Jacobian, 60, 200
matrix, 61

kinetic equation, 67, 69, 74, 75, 83, 91
Klein–Gordon equation, 163, 179
Klimontovich equation, 67–69

Lagrangian, 6, 23, 25, 26
equation, 24
Landau damping, **210**
electron, 43, 55, 109, 111, 112, 162, 193
of gravitational waves, 193
neutrino, 183
photon, 1, 6, 97, 108, 111, 112, 114, 193
Laplace transformation, 110
large-scale stochasticity, 53, 61
laser beating, 44
laser pulse,
CO_2, 5
intense, 1, 31, 43, 65, 67, 112, 140, 210, 212, 213
short, ultra-short, 1, 65, 74, 76, 78, 87, 101, 103, 137
ultraviolet, 5
linear map, 61
line broadening, 155
L-map, 57
L-mode, 64
local wavevector, 8
Lorentz,
force, 98
transformation, 17, 127, 128, 130, 132, 175
lossless medium, 9
lower hybrid frequency, 64
Lymann α line, 63

magnetic,
clouds, 57

mode, 7, 126, 129, 130, 141, 151, **210**
magnetoplasmas, 11, 63
massive vector field, 163, 165
Maxwell's equations, 69, 70, 71, 74, 83, 88, 115, 124, 125, 129
matrix transformation, 61
Mathieu equation, 139, 153
mean field acceleration, 3, 193
mean force, 85
method of the variation of parameters, 121
metric tensor, 184
microscopic density distribution, 68
microwave,
cavity, 4, 148
experiments, 36
Minkowski's metric tensor, 27
mirrors, 57
mode coupling, 7, 123, 142, 144, 148, 152
momentum conservation law, 85
mono-energetic photon beam, 113
moving,
boundaries, 15, 31, 57
magnetic field perturbations, 63

neutrino, 7, 194
dispersion relation, 177, 178, 180
effective mass, 177
fluid equations, 183
kinetic equation, 183
reflection, 181
neutron star, 178
Newtonian equation, 23
nonlinear,
coefficient, 90
current, 99, 106, 107
dynamics, 54
optical medium, 88, 210
optics, 1
phase, 90
polarization, 116
wavepacket, 99

nonlinear resonance, 52
  width, 47, 51, 53
nonstationary medium, 11
normal incidence, 124
nuclear matter, 194
number of photons, 73, 132, 167

oblique propagation, 39
obstacle epistomologique, 3
occupation numbers, 167
operator,
  creation and destruction, 160, 164, 166, 174
  displacement vector, 166
  electric field, 161, 166, 173
  energy, 165, 172
  number, 160, 167, 175
optical,
  cavity, 5, 170
  circuits, 74
  experiments, 74
  fibre, 6, 63, 87, 97, 115, 121, 193, 212
  medium, 31, 67, 193, 194
ordinary mode, 64
overlapping criterion, 53

pancake-like wavepackets, 101, 102
paradigm, 3
partially ionized gas, 37
particle accelerators, 44
phase,
  function, 9, 23
  slippage, 41
  velocity, 98
phase modulation, 193
  crossed, 5, 88, 210
  induced, 5, 31, 66, 88, **210**
  self, 5, 67, 87, **212**
phase space,
  two-dimensional, 33
  six-dimensional, 68
photon,
  acceleration, **211**

action, 23
averaged quantities, 83
bending, 186
classical concept, 8
distribution, 108
effective mass, 1, 13, 97, 114, 187, 193, **211**
effective temperature, 85
equivalent charge, 1, 2, 3, 97, 102, 103, 107, 114, 193, **211**
equivalent dipole, 6, 97, 115–122, 194, 212
gas, 6, 67
mean density, 78, 83
mean energy, 86
mean velocity, 84
number density, 70
ondulator, 6, 105, 122, 194
pair creation, 7
plasma frequency, 114
position, 10
pressure, 85
proper frequency, 185, 188
proper time, 29, 49
quantum concept, 8, 166
trajectories, 5, 42
universal frequency, 188
velocity, 8, 23, 25
photon acceleration,
  stochastic, 6
photoionization, 4, 32, 82
Planck's constant, 24
Planck distribution, 108, 112
plasma physics, 1
plasma,
  boundary, 105
  instabilities, 97
  isotropic, un-magnetized, 9, 24, 33, 75, 86, 97, 161, 211
  kinetic theory, 109
  magnetized, 11, 31, 63
  temperature, 112
  wave, 6
plasmons, 63, 165

Poincaré map, 52
Poisson bracket, 69
Poisson's equation, 48, 108
polarization states, 64, 69, 84, 160–161, 166
ponderomotive force, 2, 97, 103, 108, 115, 194, 211
pulse,
  distortion, 78
  invariant shape, 90
  steepening, 95

quantum,
  numbers, 140
  optics, 70
  oscillator, 160
  theory of collisions, 140
quasi-linear diffusion coefficient, 112
quasi-static approximation, 101

R-mode, 64
radiation,
  mechanisms, 97
  microwave, 4
  monochromatic, 6, 51, 54, 62, 87, 88
  pressure, 2, 97, 115, 183, 194, 211
  supercontinuum, 5, 212
radio galaxies, 63
ray equations, 5, 72, 82, 178, 185, 211
recombination, 32
rectangular pulse, 118, 121
reduced,
  distribution function, 111
  photon action, 23
reflection, 4, 33, 175, 193
  partial, 7, 123, 148
  total, 15, 21, 128, 132, 148
refraction, 1, 3, 12-13, 175, 193
refractive index, 9, 31, 64, 125, 193, 195

nonlinear, 4, 66, 91
relativistic,
  electron plasma waves, 103
  gamma factor, 24, 26, 27
  invariants, 127, 128, 132
  phase velocities, 43, 108, 112
  space–time, 49
relativistic mirror, 4, 17, 21, 22, 34, 148, 181
  co-propagating, 39
residual plasma frequency, 37
resonance
  asymmetry, 46
  condition, 148, 153, 155, 190
resonant,
  electrons, 44, 112
  transition, 65

saturation amplitude, 148, 150
scattering,
  Brillouin, 2
  Compton, 3, 4
  Raman, 2, 115
  Rayleigh, 3
Schwarzschild metric, 191
secondary,
  field, 116, 118
  photons, 107
self blueshift, 82, 87, **212**
self-coupling coefficient, 143, 146, 153
self-frequency shift, 32, 67
separatrix, 36–37, 46, 47
  destruction, 53
shocks, 31, 57
simple pendulum, 45
sine differential operator, 71, 74
slingshots, 3
Snell's law, 12, 14, 15, 212
  generalized, 6, 17–19, 21
soliton propagation, 118
sound wave, 137
space–time refraction, 15–17
special relativity, 27

spectroscopy, 51
spectral,
  asymmetry, 95
  broadening, 65
  undulations, 83
  width, 79, 87
spin one, 165
squeezed states, 172
standard map, 61
stars, 185, 186, 187, 191
static,
  electric field, 133, 135, 136
  magnetic field, 63
Stark effect, 65
stationary medium, 11
stochastic,
  acceleration, 6, 51
  trajectories, 156
Stokes sideband, 95
stroboscopic plot, 52
sub-cycle pulses, 137
successive reflections, 40, 43
super-Hamiltonian, 29
supernova explosion, 177
supraluminous boundaries, 19, 20
susceptibility, 9, 26, 70, 118, 196,
    199, 211
  electron, 111, 112, 113, 114
  linear, 88
  nonlinear, 66
  photon, 111, 112, 113, 114
  second-order, 116
  third-order, 90, 117

theory of field ionization, 83
threshold criterion, 62
time,
  boundary, 169, 175
  continuity conditions, 125
  discontinuity, 125, 165–172
  reflection, 1, 13–15, 19, 123,
      125–126, 167, 171, 175,
      193, **212**

refraction, 1, 13–15, 123, 193,
    212
time-dependent medium, 11
timescales, 10
topological transition, 61
transition probabilities, 65
transition radiation, 3, 105, 194
  nonlinear, 107
transform limited pulse, 77, 87
transverse electromagnetic waves,
    9, 24
trapping, 6
tunnelling ionization, 82
turning point, 34, 42
two wakefields, 52
types of trajectories, 34–38

upper-hybrid frequency, 64
uncertainty,
  principle, 87
  relations, 77
underdense fronts, 41
unlimited frequency shift, 17
unstable solutions, 153

vacuum, 9, 180, 184, 188, 194, 211
  magnetic permeability, 184
  permittivity, 9, 161, 178
  photon creation, 126, 157, 170,
      175, 176
  symmetric, 168, 169
  states, 167, 168
variational principle, 23, 28
  covariant formulation, 29
virtual photons, 135
visible light, 65

wakefield, 31, 43, 102, 109, 112,
    152, 210, **213**
  wavevector, 45
wave absorption, 64
wave equation, 70, 74, 88, 98, 103,
    106, 115, 134, 137, 179
wavefunction, 179

waveguide, 4, 9, 25
wavenumber, 9
  averaged, 78
wavepacket, 8, 54, 184
  centroid, 10
  envelope, 100
  internal structure, 67
wave-particle dualism, 8
white light, 6, 51, 62, 87, 88
Wigner function, 67, 70–73, 197,
    198, 201

quantum, 70
  reduced, 71
Wigner-Moyal equation, 71, 72, 74,
    75, 195–202
WKB solutions, 143, 144, 181
wronskian, 121

zero,
  energy, 7
  frequency photons, 209, 211